The Future of Safe Systems

Related Titles

Addressing Systems Safety Challenges
Proceedings of the Twenty-second Safety-critical Systems Symposium, Brighton, UK, 2014
Dale and Anderson (Eds)
978-1491263648

Engineering Systems for Safety
Proceedings of the Twenty-third Safety-critical Systems Symposium, Bristol, UK, 2015
Parsons and Anderson (Eds)
978-1505689082

Developing Safe Systems
Proceedings of the Twenty-fourth Safety-critical Systems Symposium, Brighton, UK, 2016
Parsons and Anderson (Eds)
978-1519420077

Developments in System Safety Engineering
Proceedings of the Twenty-fifth Safety-critical Systems Symposium, Bristol, UK, 2017
Parsons and Kelly (Eds)
978-1540796288

Evolution of System Safety
Proceedings of the Twenty-sixth Safety-critical Systems Symposium, York, UK, 2018
Parsons and Kelly (Eds)
978-1979733618

Engineering Safe Autonomy
Proceedings of the twenty-seventh Safety-Critical Systems Symposium, Bristol, UK, 2019, SCSC-150
Parsons and Kelly (Eds)
978-1729361764

Assuring Safe Autonomy
Proceedings of the twenty-eighth Safety-Critical Systems Symposium, York, UK, 2020
SCSC-154
Parsons and Nicholson (Eds)
978-1713305668

Systems and Covid-19
Proceedings of the twenty-ninth Safety-Critical Systems Symposium, Online, 2021
SCSC-161
Parsons and Nicholson (Eds)
979-8588665049

Safer Systems: The Next 30 Years
Proceedings of the thirtieth Safety-Critical Systems Symposium, Bristol, UK, 2022, SCSC-170
Parsons and Nicholson (Eds)
9798778289932

Mike Parsons
Editor

The Future of Safe Systems

Proceedings of the 31st
Safety-Critical Systems Symposium
(SSS'23)
7-9th February 2023

SCSC-179

Editor

Mike Parsons
SCSC
Department of Computer Science
University of York
Deramore Lane, York
YO10 5NG
United Kingdom

While the authors and the publishers have used reasonable endeavours to ensure that the information and guidance given in this work is correct, all parties must rely on their own skill and judgement when making use of this work and obtain professional or specialist advice before taking, or refraining from, any action on the basis of the content of this work. Neither the authors nor the publishers make any representations or warranties of any kind, express or implied, about the completeness, accuracy, reliability, suitability or availability with respect to such information and guidance for any purpose, and they will not be liable for any loss or damage including without limitation, indirect or consequential loss or damage, or any loss or damage whatsoever (including as a result of negligence) arising out of, or in connection with, the use of this work. The views and opinions expressed in this publication are those of the authors and do not necessarily reflect those of their employers, the SCSC or other organisations.

ISBN: 9798363385520

Published by the Safety-Critical Systems Club 2023.

Individual chapters © as shown on respective first pages in footnote.

All other text © Safety-Critical Systems Club C.I.C.

Cover design by Alex King

Download (free to SCSC members): scsc.uk/scsc-179

Hardcopy available on Amazon

Preface

Our world is subject to dramatic change: the war in Ukraine, climate change, the threat of new infectious diseases and a new monarch in the UK. Life is having to adjust to this new normal of 'shock' changes. The new applications in system safety are no less profound: air taxis, offshore grids, remote air traffic control centres, UK space launches, virtual hospital wards, battery-electric trains, self-driving vehicles ... the list is extensive and growing.

System safety practices and approaches must adapt to deal with these new applications and new technologies, especially in areas related to Artificial Intelligence and Machine Learning: object recognition and autonomous decision making, so crucial to new air, marine and road vehicles.

Most of us now cannot understand the complexity inside systems which ensure safety, and we cannot sensibly be relied upon to take over if the systems fail suddenly. Hence systems need to be both fail-safe, with multiple levels of resilience, and to be able to explain their decisions. Justification is everything.

Our horizons are expanding: everything we do now should be considered for impact on the environment, carbon release, wider human health and personal well-being.

The SCSC Working Groups have evolved and expanded to cover some of these new areas such as Safety of the Environment. The existing groups have all continued throughout the year and their work was presented in the form of posters this year.

As always, this forward-looking symposium addressed some of these new challenges and presented work towards practical and industrially-relevant solutions.

The Safety-Critical Systems Club has operated in support of the safety community since 1991. This 31st Club Symposium was held at the Principal York Hotel in the UK in February 2023

SSS'23 spanned three days of presentations, grouped into key themes. The keynote speakers were: Tom Anderson, University of Newcastle; Tom Hughes, EDF, Gill Kernick, Arup, John McDermid, AAIP; Catherine Menon, University of Hertfordshire, James McFadden, VCA and David Slater, Cardiff University and Cambrensis Ltd. Les Hatton from Oakwood Computing Associates gave a very entertaining after-dinner talk on the Wednesday evening.

Roger Rivett hosted a session based on the 'Room 101' TV programme on the Tuesday evening where some annoying things about system safety were forever banished. The panel comprised Tom Anderson, Nikita Johnson and Tim Kelly.

The SCSC is very grateful to all those who helped organise the event within the club and are grateful for the support of the University of York.

Mike Parsons

SCSC Mission and Aims

Mission

To promote practical systems approaches to safety for technological solutions in the real world.

> Where *"systems approaches"* is the application of analysis tools, models and methods which consider the whole system and its components; *"system"* means the whole socio-technical system in which the solution operates, including organisational culture, structure and governance, and *"technological solutions"* includes products, systems and services and combinations thereof.

Aims

1. **To build and foster an active and inclusive community of safety stakeholders:**
   a. "safety stakeholders" include practitioners (in safety specialisms and other disciplines involved in the whole lifecycle of safety related systems), managers, researchers, and those involved in governance (including policy makers, law makers, regulators and auditors)
   b. from across industry sectors, including new and non-traditional areas
   c. recognising the importance of including and nurturing early career practitioners
   d. working to remove barriers to inclusion in the community
2. **To support sharing of systems approaches to safety:**
   a. enabling wider application
   b. supporting continuing professional development
   c. encouraging interaction between early career and experienced practitioners
   d. using a variety of communication media and techniques to maximise coverage
   e. highlighting the lessons which can be learned from past experience
3. **To produce consistent guidance for safety stakeholders where not already available.**
   a. "consistent" meaning the guidance is consistent within itself, and with other guidance provided by SCSC; although SCSC will also aim to co-ordinate with external guidance this is more difficult to achieve
4. **To influence relevant standards, guidance and other publications.**
5. **To work with relevant organisations to provide a co-ordinated approach to system safety.**
6. **To minimise our environmental impact wherever possible.**

The Safety-Critical Systems Club

organiser of the

Safety-Critical Systems Symposium

Avoiding Systems-Related Harm

This is a very exciting and also a very critical time for the discipline of system safety. There are multiple new applications coming on stream, including self-driving vehicles and delivery drones, which will rely on autonomous functions. These new systems utilise technologies such as machine learning and AI; approaches where assurance is not yet established. In addition, the environments in which these systems are deployed are more fragile and protected than ever before. The SCSC has stepped up with working groups looking at such systems, and also the environment in which they operate, with a mission to produce practical guidance. This is the mission of the SCSC: using systems approaches to prevent harm.

We should not forget that many safety-critical systems in everyday life work as expected day-in day-out. Safety of these systems is accepted as routine: airbags in vehicles, air-traffic control and infusion pumps are some of the critical systems in use, on which life and property depend.

That safety-critical systems and services do work is because of the expertise and diligence of professional engineers, regulators, auditors and other practitioners. Their efforts prevent untold deaths and injuries every year. The Safety-Critical Systems Club (SCSC) has been actively engaged for over 30 years to help to ensure that this is the case, and to provide a "home" for safety professionals.

What does the SCSC do?

The SCSC maintains a website (thescsc.org, scsc.uk), which includes a diary of events, working group areas and club publications. It produces a regular newsletter, *Safety Systems*, three times a year and also a peer-reviewed e-journal published twice a year. It organises seminars, workshops and training on general safety matters or specific subjects of current concern.

Since 1993 it has organised the annual Safety-Critical Systems Symposium (SSS) where leaders in different aspects of safety from different industries, including consultants, regulators and academics, meet to exchange information and experience, with the content published in this proceedings volume.

The SCSC supports industry working groups. Currently there are active groups covering the areas of: Assurance Cases, Autonomous Systems, Data

Safety, Multicore and Manycore, Risk Ontology, Security Informed Safety, Service Assurance, Systems Approach to Safety of the Environment and Safety Culture. These working groups provide a focus for discussions within industry and produce new guidance materials.

The SCSC carries out all these activities to support its mission:

> ... To promote practical systems approaches to safety for technological solutions in the real world....

Origins

The SCSC began its work in 1991, supported by the UK Department of Trade and Industry and the Engineering and Physical Sciences Research Council. The Club has been self-sufficient since 1994. In 2021 it became a separate legal entity.

Membership

Membership may be either corporate or individual. Membership gives full web site access, the hardcopy newsletter, other mailings, and discounted entry to seminars, workshops and the annual Symposium. Membership is often paid by employers.

Corporate membership is for organisations that would like several employees to take advantage of the benefits of the SCSC. Different arrangements and packages are available. Contact alex.king@scsc.uk for more details.

There is also a short-term Publications Pass which, at very low cost, gives a month's access to all SCSC publications for non-members.

More information can be obtained at: scsc.uk/membership

Club Positions

The current and previous holders of the club positions are as follows (past holders in *italics*):

Director
Mike Parsons   2019-
Tim Kelly   *2016-2019*
Tom Anderson   *1991-2016*

Newsletter Editor
Paul Hampton   2019-
Katrina Attwood   *2016-2019*
Felix Redmill   *1991-2016*

eJournal Editor
John Spriggs   2021-

Website Editor
Brian Jepson   2004-

Steering Group Chair
Roger Rivett   2019-
Jane Fenn (deputy)   2019-
Graham Jolliffe   *2014-2019*
Brian Jepson   *2007-2014*
Bob Malcolm   *1991-2007*

University of York Coordinator
Mark Nicholson   2019-

Coordinator/Events Coordinator/Programme Coordinator
Mike Parsons   2014-
Chris Dale   *2008-2014*
Felix Redmill   *1991-2008*

Safety Futures Initiative Lead
Zoe Garstang (joint) 2020-
Nikita Johnson (joint) 2023-
Nikita Johnson   *2020-2021*

Manager
Alex King   2019-

Administrator
Alex King   2016-
Joan Atkinson   *1991-2016*

Current Working Group Leaders
Assurance Cases   Phil Williams
Autonomous Systems   Philippa Ryan
Covid-19   Peter Ladkin
Data Safety Initiative   Mike Parsons
Multicore and Manycore   Lee Jacques
Risk Ontology   Dave Banham
Safety Culture   Michael Wright
Security Informed Safety   Stephen Bull
Service Assurance   Mike Parsons
Systems Approach to Safety of the Environment   Sarah Carrington

SCSC Organisation

The SCSC is a "Community Interest Company" (CIC). A CIC is a special type of limited company which exists to benefit a community rather than private shareholders. The SCSC has:

- A 'community interest statement', explaining our plans;
- An 'asset lock'- a legal promise stating that our assets will only be used for our social objectives;
- A constitution, and
- Approval by the Regulator of Community Interest Companies.

Our community is that of Safety Practitioners. As a distinct legal entity the SCSC has more freedom and can legitimately do things such as make agreements with other bodies and own copyright on documents.

There is no change to the way we do things, or the membership we serve.

Our Company Number is 13084663.

Papers

In first named author order. See also **Abstracts** on page xv, **Author Index** at page 363 and **Title Index** on page xvii.

What the first 100 years of the railways can teach us about the first 10 years of self-driving road vehicles
Tom Anderson, Mike Parsons and Roger Rivett 21

Making the safety standards work for you
Stephen Bull, Lucia Capogna, Rob Davies, Stephen Gill 57

Case Study Analysis of STPA on an Industrial Cooperative Robot and an Autonomous Mobile Robot
Laure Buysse, Simon Whiteley, Jens Vankeirsbilck, Dries Vanoost, Jeroen Boydens, Davy Pissoort 83

A Service Analysis of the Mont Blanc Tunnel Fire
James Catmur, Kevin King, Mike Parsons, Fathi Tarada 107

Challenges in Safety Mitigation and Assurance of Multi-agent Complex Systems
Alastair Faulkner, Mark Nicholson 143

Architecting Safer Autonomous Aviation Systems
Jane Fenn, Mark Nicholson, Ganesh Pai, Michael Wilkinson 163

Grasping the Chalice – The Quest for the "Holy Grail" of Drone Operations
Paul Hampton 189

Can software engineering methods give us better software safety standards?
James Inge 215

IEC 63187 – Tackling complexity in defence systems to ensure safety
James Inge, Phil Williams 233

Anticipating Accidents through Reasoned Simulation
Craig Innes, Andrew Ireland, Yuhui Lin, Subramanian Ramamoorthy 247

Making the Water Visible: A methodology for exploring Systemic Change
Gill Kernick 265

Machine safety conformance limitations for highly automated and autonomous heavy-duty mobile machinery
Aimée M.R. de Koning, Reza Ghabcheloo 281

Safe, Ethical and Sustainable: Framing the Argument
John A McDermid, Simon Burton, Zoe Porter 297

Introducing Autonomous Systems into Operations: How the SMS has to Change
John McDermid, Mike Parsons 317

Hierarchical Approaches to Product Cyber Security: An Automotive Case Study
Robert Oates, Aditya Deshpande 339

A Practical Approach for the automation of product safety case generation in CI Framework
Doria Ramadan 355

Abstracts

In first named author order; see also full **Author Index** at page 363 and **Title Index** on page xvii.

Engineering safety-critical systems with open-source software
Paul Albertella 19

Build, monitor, and measure your live safety case in nLoop
Carmen Cârlan 105

Legal Liability for Safety Critical Systems
Dai Davis 141

Co-evolving development, implementation and Operational SMS using a Digital Twin
Alastair Faulkner, Mark Nicholson 161

How Nuclear New Builds Incorporate Lessons Learned
Tom Hughes 213

The Language of Risks and the Risks of Language
Catherine Menon, Austen Rainer 335

Rising to the challenge of certifying automated vehicles
Jamie McFadden 337

Dragons, Enigmas, Treasure Islands and Zombies: An Iconography of Risks Related to Dead Data
Mike Parsons, Paul Hampton 351

An Overview of IEEE Std 1848-2020: Standard for Techniques and Measurement to Manage Functional Safety and Other Risks with Regards to Electromagnetic Disturbances
Davy Pissoort, Keith Armstrong 353

Resilience in Safety Critical Systems – Offshore FRAM
David Slater 361

Title Index

In title order. See also full **Author Index** at page 363.

A Practical Approach for the automation of product safety case generation in CI Framework 355

A Service Analysis of the Mont Blanc Tunnel Fire 107

An Overview of IEEE Std 1848-2020: Standard for Techniques and Measurement to Manage Functional Safety and Other Risks with Regards to Electromagnetic Disturbances 353

Anticipating Accidents through Reasoned Simulation 247

Architecting Safer Autonomous Aviation Systems 163

Build, monitor, and measure your live safety case in nLoop 105

Can software engineering methods give us better software safety standards? 215

Case Study Analysis of STPA on an Industrial Cooperative Robot and an Autonomous Mobile Robot 83

Challenges in Safety Mitigation and Assurance of Multi-agent Complex Systems 143

Co-evolving development, implementation and Operational SMS using a Digital Twin 161

Dragons, Enigmas, Treasure Islands and Zombies: An Iconography of Risks Related to Dead Data 351

Engineering safety-critical systems with open-source software 19

Grasping the Chalice – The Quest for the "Holy Grail" of Drone Operations 189

Hierarchical Approaches to Product Cyber Security: An Automotive Case Study 339

How Nuclear New Builds Incorporate Lessons Learned 213

IEC 63187 – Tackling complexity in defence systems to ensure safety 233
Introducing Autonomous Systems into Operations: How the SMS has to Change 317

Legal Liability for Safety Critical Systems 141

Machine safety conformance limitations for highly automated and autonomous heavy-duty mobile machinery 281

Making the safety standards work for you 57

Making the Water Visible: A methodology for exploring Systemic Change 265

Resilience in Safety Critical Systems – Offshore FRAM 361

Rising to the challenge of certifying automated vehicles 337

Safe, Ethical and Sustainable: Framing the Argument 297

The Language of Risks and the Risks of Language 335

What the first 100 years of the railways can teach us about the first 10 years of self-driving road vehicles 21

Engineering safety-critical systems with open-source software

Paul Albertella

Codethink

Abstract *Free and open-source software (FOSS) powers a huge range of critical systems and services, including internet infrastructure, cloud services and core financial systems. FOSS also dominates the realm of software development tooling, especially for DevOps and AI/ML applications. In the development of safety-critical systems, however, the use of FOSS has historically been treated as problematic, because open-source projects do not typically apply the formal software engineering processes (e.g. quality management, requirements specification and formal design) expected by safety standards.*

FOSS can be certified as part of safety-critical products by establishing the 'missing' process evidence for the specific application. However, this purely 'remedial' approach to consuming FOSS can be counter-productive, especially if the 'remedies' are not shared by the product developers. It may also defeat one of the key benefits of the open-source model, which has already been demonstrated in the security domain: transparency as the basis of trust and continuous improvement. Furthermore, the use of increasingly complex, software-intensive systems in safety applications makes 'traditional' approaches to assurance increasingly difficult. Precisely specified behaviour for all execution paths may not be achievable, and much of the software involved in such systems (or in the toolchains used to construct and test them) may be pre-existing and generic, with complex or loosely-controlled supply chains for its dependencies.

We propose a new approach to consuming pre-existing software (both FOSS and proprietary) as part of safety-critical products and processes, using System Theoretic Process Analysis (STPA) to model the roles and responsibilities of software components in a given system, and to specify the system's safety goals: how components contribute to them, how they may be violated, and what constraints must be met in order to prevent or mitigate such hazards. Using such a model as a system specification can provide a basis for verifying an integrated system, including fault injection scenarios to validate the effectiveness of verification measures and safety mechanisms, as well as providing detailed safety requirements for individual components. It may also be used to analyse the risks involved in consuming specific components — including those arising from their software development processes — and to specify how these risks may be addressed or

© Paul Albertella 2023.
Published by the Safety-Critical Systems Club. All Rights Reserved.

mitigated. Open publication of such models for generic or anonymised systems could also enable organisations to collaborate in refining and validating the hazards considered and mitigations adopted for complex safety applications, such as those required for advanced automated driving solutions. This may provide a way to deliver the benefits of transparency to safety problems.

What the first 100 years of the railways can teach us about the first 10 years of self-driving road vehicles

Tom Anderson

>Newcastle University

Mike Parsons & Roger Rivett

>AAIP, University of York

Abstract *The development of the railways in Great Britain at the start of the 19th century was a time of great innovation in engineering, including infrastructure and rail vehicles under steam motive power. Unfortunately, it was also a time of some terrible accidents, including the fatality at the opening of the Liverpool and Manchester Railway in 1830 when Stephenson's Rocket killed William Huskisson, and the Clayton Tunnel rail crash in 1861 where confusion over signalling led to 23 fatalities when one train ran into another in the tunnel. But, most importantly, lessons were learned from these dreadful events, leading to improved designs for locomotives and rolling stock, safer methods of working and operation, and new regulations and legislation. We are now at the dawn of autonomous road transport which offers the prospect of major benefits, not least in terms of a reduction in the number of road accidents and consequential fatalities. However, autonomous vehicles incorporate critical new technologies, such as machine learning, which could interact with unforeseen scenarios and manually driven traffic in unpredictable ways. This does mean that some residual accidents are inevitable. This paper considers a selection of railway accidents from the steam age and interprets them for the 'autonomous age', reading across to show how they can still be relevant and instructive.*

Abbreviations and Acronyms

AEB	Automatic Emergency Braking
AI	Artificial Intelligence
AS	Autonomous System
ASIL	Automotive Safety Integrity Level

© Tom Anderson, Mike Parsons & Roger Rivett 2023.
Published by the Safety-Critical Systems Club. All Rights Reserved

AV	Autonomous Vehicle (refers to both conventional vehicles with autonomous features and fully autonomous vehicles)
CCTV	Closed-Circuit Television
FDIS	Final Draft International Standard
GWR	Great Western Railway
IET	Institution of Engineering and Technology
ISO	International Organization for Standardization
MP	Member of Parliament
mph	miles per hour
MRM	Minimum Risk Manoeuvre
RSSB	Rail Safety and Standards Board Ltd
V2V	Vehicle to Vehicle (communication)
V2X	Vehicle to Everything (communication)
VIP	Very Important Person
WW II	Second World War

1 Introduction

During the early years of the development of railways the rush to invest in and enlarge the railway network stimulated numerous technological advances in construction methods for track and vehicles, in the design of passenger carriages and goods wagons, in the construction of locomotives powered by steam, and in the control and signalling of traffic on the rail network.

But there were very many accidents on the growing railway system, with consequent deaths and serious injuries. In the UK this eventually led to the notion of Board of Trade appointed Railway Inspectors who reported on all substantive incidents, making recommendations for improved ways of working that could reduce the risk of further accidents thereafter. These reports and recommendations constitute a trove of insight into the causes of rail accidents, which we will attempt to plunder for lessons applicable more widely in transportation – and especially with respect to the development of autonomous road vehicles (AVs)[1]. Current practice in the UK commendably maintains the tradition of issuing safety bulletins and reporting on lessons learned from incidents (NetworkRail, 2022), while the RSSB curates a health and safety strategy for the railway (Morse, 2020), accompanied by quarterly progress reports.

Of course, lessons learned from accidents are not the only 'tool in the box'; legislation can also play a part in reducing risks and it is instructive to see how regulation and laws followed in the wake of major or repeated railway accidents

---

[1] In the 19th century the railways, and indeed wider society, lived and worked with a very different safety culture to that obtaining now, in the 21st century. It should be noted that the current development of AVs does have safety as a prime consideration, at least for traditional car manufacturers and certainly in the UK. We acknowledge the current AV focus on safety and encourage it to continue maturing.

(albeit sometimes rather slowly). It could be a valuable exercise to draw parallels between the development of early rules and regulations imposed on the railways, and the benefits (and drawbacks) of corresponding regulation for AVs – we leave this as work for future consideration, but a brief summary of the current situation is presented in section 3.

1.1 Motivation

Our aim is to recount several of the more informative and significant accidents occurring on the railways of the UK, covering the period of around 100 years starting from 1830. From each of these we will endeavour to draw out any lessons for safety that are useful and instructive for a more modern era, with a particular focus on relevance to the proposed introduction of increasingly autonomous road vehicles – the possibility of driverless cars (and trucks, and coaches, …).

We have attempted to select a set of accidents that cover a range of different incident scenarios, and with a variety of causes. Many of these accidents excited widespread interest at the time, with important recommendations from the Railway Inspector, and – in due course – legislative regulations imposed on the railway companies.

1.2 Accidents Considered

The accidents listed below were chosen as instructive examples from a larger set; there are very many more that could have been included[2]. The figures for deaths and injuries are incorporated in this listing as a reminder that these incidents were all appalling tragedies at the time; it would be improper, and surely impossible, to treat them without due regard for their horrific consequences.

1. **1830 Parkside** (Rainhilltrials, 2022) where MP William Huskisson's left leg was crushed by the *Rocket* locomotive, and he died a few hours later.
2. **1849 Plympton** (Railways Archive, 2022b) where collapse of a locomotive firebox was caused by excessive boiler pressure, most likely due to interfering with the safety valves. One death and one injury.
3. **1857 Lewisham** (Wikipedia, 2022a) where a following train ran into a train standing at a red signal just east of Lewisham station, killing 11 and injuring 30.
4. **1860 Tottenham** (Railways Archive, 2022c) where an express train was derailed at Tottenham (now Tottenham Hale) station due to a broken tyre; seven died and nine were injured.
5. **1861 Clayton Tunnel** (Wikipedia, 2022b) where one train ran into another that was reversing inside the tunnel, killing 23 and injuring 176.

---

[2] The Railways Archive website (Railways Archive, 2022a) lists over 9000 accidents on railways in the UK, as of August 2022.

6. **1862 Winchburgh** (Wikipedia, 2022c) where two trains met head-on when only one line was in use due to maintenance on the Edinburgh and Glasgow Railway. Seventeen were killed and 35 injured.
7. **1874 Thorpe St Andrew** (Gingell, 2022, Railways Archive, 2022d) where a momentary lapse of concentration combined with an inadequate protocol saw two trains come together in a catastrophic collision on single track east of Norwich, resulting in the deaths of 27, plus 73 injured.
8. **1876 Abbots Ripton** (Wikipedia, 2022d) where an express from Edinburgh to London collided with a coal train during a blizzard – a northbound express then ran into the wreckage. Fourteen killed and 58 injured.
9. **1889 Armagh** (Wikipedia, 2022e) where a crowded excursion train stalled on a steep incline; the crew then divided the train, which left the rear portion inadequately braked – those coaches ran back downhill, colliding with a following train. Eighty were killed and 260 were injured.
10. **1898 Wellingborough** (Wikipedia, 2022f) where a luggage trolley fell onto the tracks resulting in the derailment of a passenger train. Seven deaths and 65 injured.
11. **1909 Cardiff** (Railways Archive, 2022e) where a boiler explosion at the engine shed killed three and injured three more.
12. **1910 Hawes Junction** (Wikipedia, 2022g) where a double-headed night express passenger train ran into the rear of a pair of light engines in bleak moorland, causing a major fire that resulted in 12 deaths and injuries to 30.
13. **1915 Quintinshill** (Wikipedia, 2022h) where a packed war-time troop train heading south ran into a stationary local passenger train; a northbound express then collided with the wreckage to create the worst disaster on railways in the UK. The resulting death count exceeded 200 and almost 250 were injured.
14. **1921 Abermule** (Wikipedia, 2022i) where two trains collided head-on on a single-track section, both drivers believing that they had the token for that section. Seventeen died (including a director of the railway) and 36 were injured.
15. **1940 Norton Fitzwarren** (Wikipedia, 2022j) where the driver of an express running on the relief line believed he was on the main line, and so ran through the buffer stop where the relief line ended. The death toll was 27 with 56 injured.

1.3 Sources

Anyone seeking a broad background of knowledge regarding railway accidents in the UK will be well served by the book originally authored by LTC Rolt (Rolt and Kichenside, 1986). Excellent alternative accounts are available; those we consulted are included in our reference list (Earnshaw, 1996; Hall, 2006; Nock,

1970; Prosser and Keay, 2019). Supplementing these printed resources, we accessed railway accident reports from Wikipedia (2022a-j), and the extensive materials available at the website of the Railways Archive (2022a-e). Recommendations for further reading on rail accidents can also be found at the Railways Archive (2022f). A range of UK government material on AVs can be accessed via Autonomous road vehicles (2022).

1.4 Organisation

For each of the 15 selected accidents, the next section provides a condensed overview of the incident and its causes, followed by bullet point listings where we first note any recommendations for railway operation that were made at the time of the accident, and then a read-across identifying lessons or issues of ongoing relevance for safety in general, and in particular for the modern AV setting.

Section 3 refers to the enactment of legislation to regulate the railway companies and comments on the prospects for regulations controlling the development and deployment of AVs.

In section 4 we present our attempt to distill the primary concerns for design and deployment of AVs that our examination of railway accidents has identified. Clearly these are not the only issues of concern, and we do not suggest that they are being ignored by those who are closely involved with the development of AVs. Our objective is simply to reflect on the history of innovation in the railway domain, recognising the loss of life that that entailed (and this on a much larger scale than determined by our brief list of 15 accidents in just the UK), and to remind our readership of factors which are likely to remain pertinent in this modern – and rather more enlightened, at least in terms of safety – era. We also refer (in section 5) to some possibilities for extending this line of investigation.

2 Accident Descriptions and Analysis

Our overviews of these 15 accidents are deliberately concise, and some detailed facets have often been omitted – although we have, of course, aimed for an accurate portrayal of the primary circumstances. And, in a very few cases, the records that we have consulted for an event have been either incomplete, or ambiguous, or even contradictory. However, we are not aiming to provide definitive accounts of accidents on the railway; readers who seek further details are referred to our sources as a starting point.

1. Parkside 1830. This illustrates that mission requirements do not necessarily include safety provision. The choice of motive power for the nearly completed Liverpool and Manchester Railway was determined by trials of steam locomotive performance, held at Rainhill in 1829. The trial requirements were set out in meticulous detail, and concentrated on power, speed and the ability to complete the journey from Manchester to Liverpool and back.

Just under a year later the railway held its ceremonial opening. At Parkside station the Member of Parliament for Liverpool, William Huskisson, walked on the track to greet the Duke of Wellington (famous victor at Waterloo and the then Prime Minister). A now famous locomotive, Stephenson's *Rocket*, was already in motion and although Huskisson (who had reduced mobility) was seen by the driver he was unable to stop and ran the MP down, thereby causing injuries from which he died later that day – the first fatality resulting from an accident on a passenger railway. It should be noted that the *Rocket* itself was used to take the injured MP to a place where he could be treated. [1 death]

The Rainhill trials had placed no emphasis on the ability of locomotives to stop; indeed, more than 45 years after an MP was killed on the railway, steam locomotives specifically designed for high-speed running were being built with no brakes at all on the locomotive[3].

Fig.1. Taking in Water at Parkside (the station where Huskisson fell), 1831, Henry Pyall
(Public domain, via Wikimedia Commons)
(The original watercolour by T.T. Bury is in the National Railway Museum)

---

[3] For example, the famous Stirling 8' Singles, which only had a hand-screw operated brake on the tender.

Original Recommendations – none found

Read Across:
- Autonomous vehicles need to recognise the possibility of careless or even foolish behaviour from others nearby[4]
- New technology creates new hazards, which naïve users may not appreciate, understand or expect
- Some users may ignore safety precautions and think they know better
- VIPs may assume they are invulnerable to risk
- Users with reduced mobility may need special arrangements to fully benefit from new methods of transport
- AVs should be capable of recognizing (and acknowledging) when they have caused or contributed to an accident – especially if indirectly
- AVs may need to take appropriate action if they are involved in an accident, even indirectly (e.g. slow down, stop, pull over, or other MRM[5])
- Some vehicles already have a call system for the emergency services when involved in an accident. This could be extended and enhanced, for instance to detect accidents nearby, and include a 'take me to nearest hospital' mode which is enabled after an accident.

2. Plympton 1849. This illustrates how a fundamental safety feature can be compromised by misuse. The early years of steam locomotion involved all too many disasters caused by excessive steam pressure – leading to catastrophic and explosive failure of the boiler. To provide protection, boiler pressure was restricted by safety valves (usually two) designed to lift and release steam at a pre-set limit.

The incident summarised here occurred on the approach to Hemerdon Bank, a four-kilometre incline of 1 in 42 near Plymouth. This is a steep gradient for a railway, which thus poses a significant challenge to the driver and his locomotive.

A broad-gauge goods locomotive (named Goliah [sic]) blew up at the start of the gradient. The driver was thrown from the footplate, but the fireman was killed. The accident inspector determined that the second Salter safety valve had been screwed down to enable boiler pressure to reach 150 pounds/square inch before lifting, rather than the normal working pressure of 70 pounds/square inch; it would then be possible for the driver to manually hold down the first valve,

---

[4] The main change that comes with automotive autonomy is that the understanding of the current status of the external world, the prediction of its imminent changes, and the decisions made as a result of these factors are now performed by a machine rather than by a human.

[5] In practice, the choice of MRM at any given time is limited by the technology available on the vehicle, the nature of the vehicle failure (if present) and the situation. It is merely the least-worst of the available options and does not guarantee the avoidance of accident, injury or fatality.

thereby increasing the power of the engine – but at gravely increased risk[6]. [1 death; 1 injured]

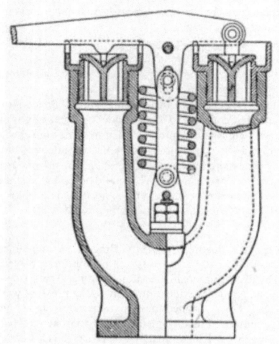

Fig. 2. Ramsbottom safety valve
(Andy Dingley (scanner), Public domain, via Wikimedia Commons)

In 1856 the Ramsbottom safety valve was invented in order to make any such tampering much more difficult, greatly reducing (but not eliminating, as we shall see in 1909 at Cardiff) the occurrence of boiler explosions.

Original Recommendations:
- Safety valves to be regularly checked (weekly was specified for the GWR)
- Limits on boiler pressure must be strictly adhered to

---

[6] The inspector appears to have accepted the driver's claim that the first valve was *not* held down, despite a witness claim that it was. It is perhaps indicative that a week after the accident the Locomotive Superintendent (D. Gooch) issued an edict strictly forbidding any improper adjustments to safety valve settings.

Read Across:
- Safety measures and protection devices are rooted in necessity, and are not to be treated as obstacles that can be evaded
- There may be occasions when safety features are disabled or 'gamed' for some perceived advantage
- Some drivers, garages, enthusiasts or hackers may want to 'improve' (as they will see it) the manufacturer supplied characteristics of AVs, usually related to performance
- Control software and data may be targeted for unauthorised modification (and hardware too)

3. Lewisham 1857. This illustrates how procedures are not always followed and subsequent cover-ups are possible. A late running passenger train was stopped and held at a red signal just to the east of Lewisham station. The train guard heard the following train approaching and ran back towards it, waving a red lamp and blowing a whistle, to no avail. A collision ensued, at about 20 mph. The second train should not have left Blackheath, the preceding station, since a "line clear" telegraph message had not been sent (from Lewisham); however, the Blackheath signalman had nonetheless recorded that the line clear message had been received. [11 deaths, 30 injured]

Original Recommendations:
- Improvements to the operational procedure and recording of telegraph communication, and coordination with line-side signalling
- Adherence to the company rule book as regards signalling
- Improvements to passenger train braking

Read Across:
- Operators (e.g. in the case of a fleet of AVs) or owners may abuse technological safety features
- Operators or owners of AVs (or, indeed, operators/maintainers of road infrastructure) may falsify records to avoid blame
- Communications, logs and records relating to an accident need to be trustworthy and preserved (especially across different manufacturers)
- The establishment of an independent authoritative accident investigation body for incidents involving AVs would be beneficial[7]

---

[7] Public roads do not have a centralised operator (at least in the UK), as there is with the railways. Each vehicle is its own master and there is no overall authority responsible for signals, vehicle movement, etc. This is a significant challenge in that AVs do not benefit from a level of supervision as do railways. Historically this is always finessed by putting all responsibility for whatever happened on the human driver.

- Responsibility and, potentially, liability for a road accident is often debatable – and may be subject to legal disputation; the highly technical nature of the systems supporting AVs will increase the complexity of such controversies
- It is important to be able to decide when data received from infrastructure (V2X) and, critically, from other vehicles (V2V) can be relied upon

4. Tottenham 1860. This illustrates how manufacturing defects can cause accidents, even after considerable elapsed time. The tyre of a leading wheel on a locomotive entering Tottenham station broke apart due to an undetected poorly-made weld, derailing the locomotive and its following coaches, which then crashed against the platform. Railway vehicle components, construction (and maintenance) need to meet appropriate quality standards to meet safety requirements, and these must be monitored. Thanks to advances in metalworking, tyres for railway wheels are now made as cast steel rings which are shrunk on to the wheel. But back in the 19th century the tyre was made from a metal strip fitted to the wheel and then welded to form a ring. [7 deaths, 9 injured]

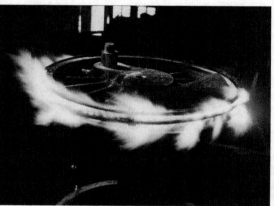

Fig. 3. Re-tyring a locomotive driving wheel
(Jack Delano, Public domain, via Wikimedia Commons)

Original Recommendation:
- The coroner's jury declared that "*had proper caution and vigilance been used the [defective weld] might have been detected*"

Read-Across:
- Manufacturing defects can go undetected, despite best manufacturer efforts

- The failure characteristics of software and data are not the same as mechanical wear – there may well be no indication whatsoever that a failure is about to happen
- Tyre failures may occur on AVs, and they need to be handled in an appropriate way – more generally there is the challenge of designing a vehicle that can cope with extreme events that arrive without warning[8]

5. Clayton Tunnel 1861. This illustrates that protocols need to be robust, cover all sequences of events, be understood and followed by all operators. Before the introduction of block working many railways operated on a time interval system (requiring at least 5 minutes between trains on the line at Clayton tunnel). Signal boxes controlled the entry and exit at each end of the tunnel and could communicate by means of a rudimentary telegraph. The south box operated a signal located 300 m before the signal box; this signal had a safety feature: when a train passed it automatically returned to danger. However, since this feature was known to be unreliable, any failure to return to danger rang an alarm bell in the signal box. The signalman also had a red flag (to indicate stop) and a white flag (to indicate all clear).

On the day of the accident three long passenger trains left Brighton station (about 8 km south of the tunnel) with time separations of only 3 then 4 minutes. The first of these trains was passed into the tunnel, but the signal did not return to danger and so the alarm sounded. The signalman was busy using the telegraph to inform the north box that a train was heading through the tunnel, and so did not immediately manually rectify the errant signal. When he did so, the second train had already passed the clear signal and had almost reached the signal box. He rushed to use his red flag, but the train continued onwards and into the tunnel. He, rather confusingly, again signalled "train in tunnel" to the north box. So now he asked the north box: "is tunnel clear?" Just then the first train emerged from the tunnel, so the north box responded positively "tunnel clear". The relieved signalman therefore removed his red flag, enabling the third train (which had slowed) to follow on into the tunnel.

Most unfortunately, the driver of the second train had glimpsed the red flag and stopped his train almost 1 km inside the tunnel. He then decided to slowly set back to the signal box for advice. The last coach of his train was about 200 m from the tunnel portal when the locomotive of the third train violently collided with it. [23 deaths, 176 injured]

Original Recommendations:
- Brighton station staff were censured for sending the three trains out onto a difficult section at such close intervals

---

[8] It has always been accepted that a human driver may not be able to cope with such events, but this may be considered unacceptable for a machine

- The signalman was working a 24-hour shift which was criticised as being clearly excessive
- Automatic warning systems which are known to be unreliable should be considered more dangerous than a simple manual device
- The limited telegraph communication employed at the tunnel was clearly inadequate; there was no retained information, yet it was apparent that all that was needed was a determination as to whether the section was occupied or clear
- A firm (and detailed) recommendation was made for the adoption of "absolute block working", to be made effective as soon as possible

Read-Across:
- Safety features can create new hazards, which may not be obvious
- When unreliable communication mechanisms are unavoidable they need to be supplemented with alternative reliable and fail-safe mechanisms
- Sensors and indicators should be able to detect their own faults and notify or alert as appropriate (note that a human occupant of an AV may have no understanding of what a particular sensor failure means for the vehicle; this information should be communicated to the manufacturer or maintainer)
- Driver (and passenger) actions should be evaluated and possibly challenged (e.g. in moments of panic a driver may take back manual control and make the situation worse)[9]
- Multiple lines of defence are required for critical situations or operations (possibly also involving external infrastructure, e.g. overhead warning signs to slow down could be supplemented by V2V and V2X)
- The level of toleration of failures (of sensors, infrastructure, etc.) must be defined, for example by the use of ASIL values as described in the ISO 26262 standard
- Road tunnels present additional hazards for AVs. Would an AV be able to reverse out of a dangerous situation? How should it respond if its sensors were obscured by smoke with no radio contact (e.g. the Mont Blanc tunnel fire (Wikipedia, 2022k))? For instance, careful forward progress through the smoke/fire might be the best thing to do…

---

[9] This is a very difficult balance. If a driver rightly tries to intervene, but is first challenged before being allowed to do so, the time taken for the challenge may mean that the driver no longer has sufficient time remaining to avoid an accident. Hence the need for evaluation.

6. Winchburgh 1862. This illustrates how a maintenance situation can create new risks. A section of the double-track railway between Edinburgh and Glasgow to the west of Winchburgh station was being operated as a single line to enable track maintenance. Safety was to be ensured by a pilot engine escorting all trains through the single-track section. Unwisely, the distinctive pilot engine had been replaced by a larger locomotive which was also performing additional tasks such as moving trucks.

A passenger train from Glasgow, heading east, was being followed by a ballast train which an inexperienced pointsman took to be the pilot; he therefore allowed the passenger train to proceed. But the section was already occupied by a train from Edinburgh heading west, properly accompanied by the pilot. Both drivers were able to apply brakes so that the collision occurred at less than 30 mph. [17 deaths, 35 injured]

Fig. 4. Winchburgh rail accident of 1862
(Unknown author, CC0, via Wikimedia Commons, Railways Archive – Accidents Archives)
(http://www.railwaysarchive.co.uk/eventsummary.php?eventID=26)

Original Recommendations:
- The Inspector criticised "lax procedures" and "vague instructions"
- The pilot locomotive should have been clearly identifiable

Read-Across:
- Roadworks will present additional problems for AVs (e.g. temporary markings, manual signalling, speed limits, etc.) which may be site-specific and therefore difficult to comprehend or negotiate. Consequently, mistakes could happen because of visual confusion; this may be attributable to camera or vision systems, image processing, object recognition or other functions

- Roadworks may have additional specialist vehicles temporarily present on the carriageway which may not be recognised correctly (road-rollers, etc.)
- It cannot be assumed that the normal 'rules of the road' apply through all roadworks
- Safety procedures and protocols must be clear, simple and strictly adhered to

7. Thorpe St Andrew 1874. This accident provides an example of how an operating protocol that was safe in principle was all too easily subverted by human error. In autumn 1874 the railway east from Norwich Thorpe station to Brundall was still being worked as a single-track line (in fact, it was almost ready to be upgraded to double-track). The protocol for safe operation over the single-track section was to adhere to a predefined sequencing schedule, with exceptions to be approved only by formal authorisations between the endpoints using telegraphy. Despite the need for regular exceptions (usually due to late running) the protocol had been used without mishap for over 25 years.

On the evening of the accident a mail train was waiting at Brundall because the express from London was late and had not yet arrived at Norwich. The normal schedule would have first sent the express to Brundall, and then brought the mail train through to Norwich. The Norwich station inspector wanted to make an exception and bring up the mail train; his boss, the station master, disagreed. Their discussion ended with the station master declaring "All right!", which could be taken either way. However, the inspector decided that he had approval for authorising the mail train to come up, and did so via the Norwich telegraph clerk. The young (age 17) clerk obeyed the inspector's instruction, formally recorded the authorisation as required and telegraphed the message to Brundall; however, he did not obtain (as the rules required) the inspector's signature. At Brundall the station master's son (age 12) received and acknowledged this message, and the mail train was sent on. The express arrived at Norwich just a minute later, and a new locomotive was coupled to the rear (Norwich Thorpe station is a terminus).

After a brief conversation beside the locomotive, the inspector incredibly confirmed that the express was authorised to proceed, and so it set off towards Brundall, leading to the inevitable collision. The inspector and the clerk were both at fault, and both were prosecuted. At the subsequent trial much of the oral evidence was highly contested; the inspector was convicted and the clerk acquitted. [27 deaths, 75 injured]

Original Recommendations:

- It was noted that having a unique token for the single-track section would have provided a very high level of protection (at the cost of less flexibility)[10]
- That whatever system is in place to achieve safety it should be reinforced by *"safeguards precisely expressed and carefully observed"*
- It was suggested that the best safety systems are those that can only fail when several persons simultaneously and glaringly disregard the safeguards

Read-Across:
- Process and procedure are often abbreviated (to save time or money, or when developers or operators are under pressure). This may be an issue for AV manufacturers hurrying to get a new model ready to market, for vehicle maintenance, or for fleet operators
- Lack of communication can result in fatalities. There was *"no way of communicating with either driver or stopping the trains"*. In the AV context communications could be critical, for example in convoys of AVs
- When blame is to be apportioned, with a risk of heavy penalties, there is a likelihood of individuals dissembling to evade guilt: *"In giving evidence, both men tried to shift the blame onto one another"*; after an accident involving an AV the evidence supplied by individuals may also be in conflict with their (lack of) understanding of the behaviour of the AV
- Although in this incident no blame attaches to the boy in the telegraph office at Brundall, there is obvious concern for the potentially dangerous actions of young people in any transport scenario
- A cautionary observation is that compensation after an accident can be considerable; the sum of £40,000 paid out in compensation for this railway accident in 1874 would now be worth just over £5.3 million

---

[10] This accident east of Norwich Thorpe was the worst single-track accident ever to occur in the UK, but it prompted the invention by Edward Tyer of a much more flexible token machine, patented in 1878. A description of its operation is included in the summary of the accident at Abermule in 1921.

8. Abbots Ripton 1876. This accident illustrates the danger of situational confusion caused by inaccurate signalling. Heading slowly south from Peterborough in a severe blizzard, a full coal train was to be shunted to clear the way for the "Flying Scotsman", which was following roughly 20 minutes behind. With the heavy snow reducing the driver's sighting of signals, he ran past the red signal where he was supposed to move off the main line, and so was passed forward through three block sections to Abbots Ripton where the signalman was able to attract his attention by waving a red lamp. The coal train was then instructed to reverse into a siding and started to do so. By then the express was only nine minutes behind and was running past all signals. With six full coal wagons still on the main line the Scotsman crashed into them, the locomotive and tender falling on to the northbound track and coaches piling up against them.

The immediate requirement was to protect the blocked track. A platelayer ran south placing detonators on the northbound line. The coal train driver uncoupled his locomotive and headed slowly south with his guard waving a red light from the footplate. Unfortunately, a northbound express travelling at 50 mph encountered both warnings just 1,000 metres from the accident. Despite desperate braking the driver was unable to stop, hindered by a down gradient and with the railhead slippery with ice and snow; a second collision at about 15 mph caused more coaches to pile up on the existing wreckage. An immediate examination of signals on the southbound track confirmed that all were showing "clear" despite the signalmen having set them to "stop". Icing of a slot in the signal mechanism was preventing the signals from returning to danger; even after the slot was cleared, in some cases the weight of accumulated ice on the wire running back to the signal cabin was holding the signal at clear[11]. [14 deaths, 58 injured]

Original Recommendations:
- Heavy snowfall should be treated the same as fog; in reduced visibility, railway fog men placed detonators when signals were at danger, only removing them for the periods when the signals were clear
- Trains should not run at their normal speeds in poor weather
- Express passenger trains should have high capability braking systems
- Signals should be designed so that their default position was danger (the opposite of practice at that time)
- Signals should be designed so that any impediment in their linkage forced them to danger (the opposite of most designs at that time)

---

[11] Not until 1892 was a requirement imposed to use green as the indicator for clear (rather than the previous use of white); in the same year it was made a rule that distant and home signals must be interlocked.

Read-Across:
- Road signals may be incorrect or missing
- AVs need to be aware of weather conditions, and respond appropriately as conditions change; in particular, poor visibility
- AV sensors need to maintain calibration, noting that AVs may have sensors that are unfamiliar to traditional vehicle maintainers
- AV sensors and actuators need to work effectively throughout the full range of environmental conditions – history tells us that there will always be extreme and unexpected conditions, and so resilience is key
- Developers should incorporate the capability for an AV to respond to emergency or temporary signals given by people on the road (e.g. hand signals, warning flares, red flags etc.[12])
- Roadside balises that provide information on road conditions and incidents ahead have the potential to enhance AV safety

9. Armagh 1889. This disaster provided a telling example of the need for improved braking systems on the railway. In June 1889 a very heavy excursion train of 15 coaches left Armagh station to climb a 1:75 bank towards Newry. Although the train was vacuum braked throughout, the mechanism was "simple" rather than automatic, meaning that vacuum was used (when needed) to apply the brakes. Due to inadequate communications, the locomotive assigned to the excursion was barely powerful enough for the task, and indeed failed to reach the summit – the train came to a stand about 200 m below the top of the incline. After a debate between the driver and a supervisor (with input from one of the two guards), recognising that a restart would be impossible, it was decided (most unwisely) to divide the train, leaving just five coaches attached to the locomotive. Unwise, because the brakes on the rear ten coaches would thereby be disconnected from the train pipe providing a vacuum (maintained by the locomotive) and would be held solely by the guard's handbrake at the rear. Therefore, the train crew proposed to "scotch" the wheels of some of the coaches with stones, but this was only done to a very limited extent by the two guards (who, as well as the supervisor, were thus not actually on the train).

Uncoupling was then achieved, with the rear coaches remaining stationary. However, as the locomotive prepared to move off with the front portion of the train, it (as might have been anticipated) dropped back a little before moving forward, and this pushed the rear coaches backwards, dislodging the stones and overpowering the handbrake. Once in motion the coaches accelerated down the hill and had reached a speed of over 40 mph when they ran into the following train, which had almost stopped thanks to swift action by its driver.

---

[12] The use of red flannel petticoats to warn of a landslip on the railway has been reported [Nesbit, Edith (1906), The Railway Children, Wells, Gardner, Darton].

The locomotive of the following train was knocked onto its side, and a snapped coupling divided that train into tender plus horsebox, and then five coaches. Again, the primary vacuum brake on these two portions was rendered inoperative, but the driver (now in his tender) and the guard at the rear were able to stop both portions using their handbrakes. Soon after the accident the Regulation of Railways Act of 1889 was passed by parliament, giving power to the Board of Trade to insist on "continuous automatic" brakes on passenger trains. This terminology refers to a braking system that (i) operates on all vehicles of the train (continuous), and furthermore (ii) works by holding the brakes off, so that any failure of the system results in the brakes being applied (automatic). [80 deaths, 260 injured]

Fig. 5. Aftermath of the Armagh disaster
(Illustrated London News, June 22, 1889, Public domain, via Wikimedia Commons,
http://www.old-print.com/mas_assets/full/N1650889283.jpg)

Original Recommendation:
- The Act to require automatic braking on passenger trains was already in preparation, so the major general who conducted the inquiry simply observed that such a braking system would have prevented the disaster

Read-Across:
- Automatic braking functions within an AV could produce an improvement in safety; however, it should be noted that experience with existing systems, such as AEB in current manual vehicles, has shown that such systems are not without problem

- Communication between vehicles (V2V) and other entities such as infrastructure (V2X) might also be of benefit when braking is required
- The operation of an AV may need to take account of the driver (and other occupants) leaving the vehicle when the vehicle is stopped (e.g. in a traffic jam); in most similar situations it would be inappropriate for the temporarily unoccupied AV to, for instance, move forward to maintain a place in the traffic queue

10. Wellingborough 1898. This shows that a relatively minor obstruction can have very serious consequences. A loaded luggage trolley was left unattended in a passageway at Wellingborough station. The passage sloped towards the platform, which itself sloped gently towards the tracks. The trolley ran down the passage, crossed the platform and fell down on to the line, with the London to Manchester express due imminently. Despite desperate attempts to move the heavy trolley out of the way the express struck this unexpected obstacle on the track[13], derailing just the leading bogie wheels. However, a short distance further along, the locomotive encountered point-work, which resulted in a complete derailment, followed by the wreck of the train carriages. [7 killed, 65 injured]

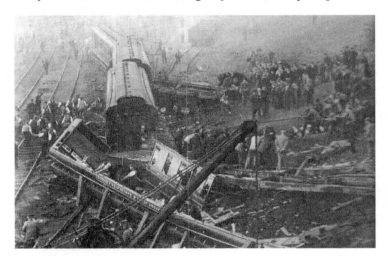

Fig. 6. Aftermath of the Accident at Wellingborough
(https://www.railwaysarchive.co.uk/eventimages.php?eventID=69&imageID=98)

---

[13] We briefly mention a more recent "obstacle on the track". At Hixon in 1968 the obstacle at a newly installed half-barrier automatic level crossing was a 120-ton transformer on a 32-wheel trailer [11 killed, 45 injured]. This led to the second Public Enquiry into a UK rail accident (the first was after the Tay Bridge disaster in 1879).

Original Recommendations:
- Platforms should not slope towards the railway tracks
- Trolleys used at stations should have brakes which were normally applied, but which could be released when the trolley needed to be moved

Read Across:
- Expect the unexpected – it is likely that something will appear in the road that has not been anticipated
- It is unreasonable to expect a machine-learning based system to be capable of positively and specifically identifying everything that might appear in the path of an AV; broad heuristic approaches will be needed
- Multiple problems may occur together, to create a 'Perfect Storm'
- Note that an AV also needs to be able to recognise when it has been involved in an accident, especially when it has caused an accident (perhaps indirectly. Such "accident awareness" is a complex problem, as the accident may have occurred some distance away (e.g. the AV has stopped too quickly, causing a shunt in a queue) or may have occurred several seconds after the AV has passed (e.g. making a cyclist wobble and eventually crash into the kerb)

11. Cardiff 1909. This illustrates how safety instrumentation can be misinterpreted, and protection mechanisms can be rendered inoperative. Steam pressure is a clear and present danger when contained by a relatively thin metal shell, the boiler of a steam locomotive. A gauge in the cab provides a direct readout of boiler pressure, and protection is provided by safety valves that release steam if pressure ever rises above a set threshold (see the discussion of the Plympton explosion in 1849).

On this occasion a driver arriving at Cardiff Docks motive power depot to take over a locomotive complained to the shed foreman that the injectors were not working and that the steam pressure gauge was not recording any pressure in the boiler. In fact, the pressure release valves had been repaired but reassembled incorrectly so that they could not open, and the pressure gauge needle had reached the zero on the wrong side (which explains why the injectors could not operate). The boiler exploded in the shed, completely destroying the locomotive. [3 deaths, 3 injured]

Original Recommendation:
- After safety valves have been repaired or modified it should be standard procedure to have the work independently inspected and then test that the valves open at the correct boiler pressure

Read Across:
- Faults and failures should be communicated to those who need to know (vehicle occupants, vehicle maintainers, etc.) in a readable and understandable way
- Updating, upgrading, bug-fixing and new releases of software will be the norm, so these updates, and the procedure for their installation, must be appropriate – this shows the importance of the ISO/FDIS 24089 standard already in development (ISO/FDIS, 2022)
- Adequate protection against all illegitimate modifications to software and data is essential to avoid these becoming a source of hazards

12. Hawes Junction 1910. An illustration of how basic procedural errors in a busy situation resulted in a major accident. The Settle and Carlisle line was built as a route for the then Midland Railway company to reach Scotland; it passes over high and rugged moorland terrain with the summit at Aisgill, 356 metres above sea level. At the beginning of the 20th century the station to the south of Aisgill was Hawes Junction (now named Garsdale). Despite its remote location, Hawes was often very busy, in part as a result of the Midland Railway's "small engine" policy which required the use of two locomotives for heavy trains on the steep gradients of the line. Consequently, there could be numerous spare engines to return north to Carlisle or south to Leeds.

On the night of the accident two engines were coupled in the branch platform track at Hawes ready to head north, while three others were being prepared to return south. An express going to Carlisle passed through the station, and the two light engines were then brought out onto the main line by the signalman, held on the station advanced starting signal awaiting clearance to follow. The signalman then concentrated on the three engines to go south and a southbound goods train, but despite receiving clearance from Aisgill he left the two Carlisle engines in place. After a few minutes of standing, their crews should have reminded the signalman of their location, but they did not do this.

Subsequently the signalman was offered and accepted a night sleeper express from the south and cleared all his signals for that express, completely forgetting the light engines, which left to start their return to Carlisle. Just 4 minutes later the double-headed express passed through, running at about 65 mph, swiftly catching up with the two engines travelling at around 25 mph. A terrific collision ensued less than 2 minutes later, about 2.5 kilometres to the north. The express train was completely derailed (bar the last coach) but the front two coaches were telescoped, rupturing the tanks holding the pressurised gas used for lighting. The resulting fire destroyed most of the coaches. [12 deaths, 30 injured]

Fig. 7. Accident at Hawes Junction on 24th December 1910 (https://www.railwaysarchive.co.uk/eventimages.php?eventID=78&imageID=426)

Original Recommendations:

- In a busy environment, occasional human errors of procedure are inevitable, so greater levels of built-in protection were deemed to be advisable
- A strong recommendation that track circuiting should be installed at Hawes to detect track occupancy, and thereby lock as necessary the relevant signal levers[14]
- A further recommendation (often repeated) was that gas cylinders should be protected with a cut-off valve in case gas lines were severed,

---

[14] Track circuit protection was invented by an American, W Robinson, who patented the technique in 1872. Although widely used on North American railroads by the start of the 20th century, there were few installations in the UK. The report on the Hawes Junction accident was a major factor in encouraging increased utilisation of this valuable safety measure – the Midland Railway reacted very positively to the inspector's recommendation and over 900 track circuits were subsequently installed on the Midland network.

but that electric lighting was much safer and so gas lighting should be phased out as soon as this could be achieved

Read-Across:
- It is a challenge for AV systems to process highly complex situations
- System overload increases the risk of a reversion to manual control (when possible) in difficult situations
- Autonomous trucks and other heavy vehicles may in future contain large hydrogen tanks with a different flammability profile to diesel, and different detection and mitigation features may be needed (especially if these vehicles are completely unmanned)

13. Quintinshill 1915. This illustrates how multiple errors in procedure can combine to create a catastrophe. The night mail from Euston to Glasgow was regularly late, and so the adjustments to accommodate this late running became accepted practice (it should have been considered exceptional, and therefore demanding more caution). On the morning of the accident, the necessary accommodation was to move a northbound local train across to the southbound track to clear the way for the express from London.

The breach of regulations by the night signaller at Quintinshill box in handing over to the day man 30 minutes after his own shift ended (seen by them as unimportant and a major convenience) was a serious breach because of the lack of concentration that it generated. Other railway staff present and gossiping in the signal box – another breach – was seen by staff as company officialdom imposing on freedoms, but generated further distraction to the responsible safety officer, the active signalman. Crucially, the failure of the night signalman to protect the local train standing outside the box by placing a locking collar on the relevant signal lever. Crucially, the failure of the night signalman to notify the box to the north that the southbound line was now occupied ("blocking back"). Crucially, the failure of the day signalman to check that a collar was locking the signal lever that was protecting the local train he had just arrived on. Crucially, the failure of the fireman of that train to check that the collar was locking the signal lever that was protecting his train, even though he signed to say that he had checked.

And then the day man (i) cleared signals to accept the northbound night mail express (fine, the road was clear); (ii) accepted and then cleared signals for a southbound troop train special (disaster, the local train was blocking the line). The troop train collided with the local and then the express ran into the wreckage. Gas from the gas-lighting cylinders on the troop train swiftly created a raging fire which burned for 24 hours. [200+ deaths, c. 250 injured]

Fig. 8. Rail accident at Quintinshill, May 22, 1915, near Gretna Green, Scotland
Extinguishing a fire in a coach
(Unknown author, public domain, via Wikimedia Commons.
Licensed under the Creative Commons Attribution-Share Alike 3.0 Unported license.)

Original Recommendations:
- The abolition of gas lighting was once again strongly urged
- Passenger rolling stock should be of all steel construction
- More fire extinguishers should be carried on passenger trains
- No criticism of the infrastructure or procedures (had they been followed correctly) was recorded, and although track circuiting would have prevented the accident it was acknowledged that Quintinshill was not a location that justified its installation

Read-Across:
- Accidents are often due to a combination of circumstances, which are often hard to see in advance, and this applies to new technology in AVs
- Accidents are often due to due a failure to adhere to protocols or procedural protections in a system (e.g. the possibility of shortcuts taken during AV maintenance)
- AVs will require additional health monitoring to identify any serious problems (e.g. overheating or fuel escape) as there will be little or no human monitoring or awareness
- Occupants will have to be alerted if there is a failure requiring them to leave the vehicle; however, instructions to occupiers to evacuate could be ignored
- It may be the case that an AV should also alert nearby vehicles – e.g. in the case of fire
- Maintainers and vehicle occupants may perform unsafe actions

- Highway service operators and maintainers monitoring the road may need to adapt warning signage to be effective for AVs (e.g. a manually held sign for roadworks may not be recognised)

14. Abermule 1921. This illustrates how excellent technology protection can be subverted by human error in a confused situation (itself partly as a result of lazy installation). The Tyer tablet instrument was a telegraphic invention to get over the limitation on a single line section that trains must alternate (one up, one down) carrying the unique permission token (the tablet) for that section. Basically, if the tablet is at one end of the section, and you want it at the other end for a following train: you place the physical tablet for the section in the Tyer instrument at one end; it notifies the instrument at the other end; the instrument at the other end can now release a physical tablet for the section. The equipment is designed to guarantee that at most one physical tablet for a section is available at any time, but once that tablet is inserted in the instrument at one location, an instance of that tablet can be released at the other location. Clearly foolproof. Well, not so at Abermule station.

A train was approaching from the north and another from the south. Both were carrying the tablet for their relevant sections. All was still well. The train from the north arrived and stopped at the station, handing over the tablet to the station junior who, on his way back to the instrument (in the station office), met the station master who asked him about the train from the south. The boy answered in a way that gave the incorrect impression (no more) that it was running late and handed over the tablet. The station master had not seen the boy collect the tablet, and mistakenly assumed that the boy had given him the tablet for the section to the south. Without looking at the tablet, the station master then took it back to the driver still waiting at the southbound platform. The driver is responsible for checking that he has the correct tablet, but he did not do so. Quoting Rolt *"Without troubling to remove it from its pouch, the driver placed it in his cab. It was his death warrant."*.

Points were set (at a separate ground frame) and signals cleared so that the train could head south where it collided head-on with the northbound express. Misguidedly, at Abermule the Tyer instruments were located in the station office; if they had rather been in the signal cabin then the signalman would have realised that a mistake had been made. [17 deaths, 36 injured]

Fig. 9. Portrait of Lord Herbert Vane-Tempest, a director of the Cambrian Railways, who died in January 1921 in the Abermule disaster
(The Graphic newspaper, public domain, via Wikimedia Commons)

Original Recommendations:
- That tablet instruments should always be located in the signal cabin under the control of the signalman, thereby minimising the risk of confused (mis)communication
- That there should always be interlocking between the tablet instrument and the signalling so that it was impossible to clear a signal for entry to a section unless the correct tablet had been issued
- A further recommendation was for the elimination of the ground frame so that all points were set from the signal cabin

Read-Across:
- Technology to detect breaches of operational due diligence (e.g. hands on wheel sensors) can be subverted
- After some time without incidents, operators can become lax and behaviours complacent (indeed, drivers of vehicles with autonomous functions may become lazy very quickly)
- Misunderstandings are likely to occur at handover points (e.g. moving to/from manual control)

- Fleet operators may not follow procedures properly (e.g. for maintenance)
- Interlocks may be broken

15. Norton Fitzwarren 1940. This illustrates that regular practices can lead to an assumption that nothing changes. Nonstandard signal placement on a dark and mirky night, with wartime blackout conditions, together with an assumption that all was as usual, created a catastrophe that could have been much, much worse.

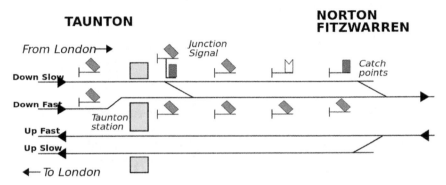

Fig. 10. A plan of the track layout between Taunton and Norton Fitzwarren indicating signal aspects showing on the down lines at the time of the 1940 crash.
(Chris McKenna (Thryduulf), CC BY 2.0 UK https://creativecommons.org/licenses/by/2.0/uk/deed.en, via Wikimedia Commons)

A heavy passenger express sleeper train travelling from London to Penzance behind a powerful "King" class locomotive had left Taunton station over an hour behind schedule in the middle of a dark wet night – particularly dark due to blackout conditions during WW II. Following down the line was a light newspaper train (also hauled by a locomotive of that same class) making good time. The railway was 4-track at that point, with both main and relief lines (in both directions) until just after Norton Fitzwarren station.

That night, the signalman routed the express onto the west bound relief line to leave the main clear for the newspaper train to overtake it, an atypical arrangement but perfectly legitimate. Just west of Taunton the relief signals were (unusually for the railway in that area, at that time) positioned on the left of the track. However, the driver of the express believed that he was, as usual on the main line, and looked only for signals to his right – these were all at green, for the newspaper train. An automatic warning in his cab that signals were being passed at danger must have been cancelled, either subconsciously or because they appeared to be contradicting the clearly visible "clear" signals; stress may have been a contributing factor, given that the driver's house had been recently damaged by bombing.

The driver realised his mistake only when the faster following train drew alongside. Immediate emergency braking could not stop his train before the relief track ended, barely 300 metres ahead, where catch points ensured that the express continued onto a short section of siding, then smashed through the buffer stops and fell onto its side in the soft ground beyond. The wreckage of the train spilled right across the main line. Most fortunately, the newspaper train had just got clear; indeed, this was so close that ballast thrown up by the crashing train scoured the last two coaches (a rivet from the locomotive actually struck the newspaper train guard). [27 killed, 56 injured]

Original Recommendation:
- That consistency in the positioning of signals was highly desirable

Read-Across:
- There is potential for accidents to result from an AV misreading or misunderstanding road signage, especially in adverse environmental conditions or poor weather
- Road signage may not always be located in standard positions (e.g. through roadworks) – sometimes signage may be missing or obscured or damaged
- A loss of situational awareness by an AV could result from a variety of abnormal situations (e.g. intervention by police) and especially when there are multiple factors in play simultaneously
- Genuine warnings or alerts produced by AV systems may be cancelled, ignored or overridden
- In exceptional conditions (e.g. in the presence of emergency vehicles) extra measures may be needed to ensure safety
- It is unclear how existing road monitoring and signalling systems (e.g. overhead gantry displays) will need to be modified for AV usage
- Drivers of AVs, when subject to stress, may seek to override the autonomous control of the vehicle, reclaiming manual control unnecessarily and making incorrect decisions that could cause harm

3 Rules, Regulations and Legislation

Many new items of legislation were passed during the early years of railway operation; in fact, 71 Acts of Parliament were passed related to railways between 1831 and 1837. A parliamentary Select Committee was established in 1839. The "Father of Railways", George Stephenson himself, gave evidence in support of better management and regulation (Hall, 2006). Three key Acts were:

- Regulation of Railways Act 1840, created the Railways Inspectorate
- Regulation of Railways Act 1873, required all companies to report on progress of installation of interlocking gear and block working
- Regulation of Railways Act 1889, imposed compulsion to adopt better braking systems

Other significant safety improvements that were recommended and then required include: replacement of gas lighting by electric, the use of steel rather than wood for passenger coaches, and better signalling practices – leading eventually to automatic alerts and warning systems.

The legal basis for the use of AVs in the UK is still being developed, but the law commission has made some key recommendations (Lawcom.gov.uk):

- Writing the test for self-driving into law (with a clear distinction made between full automation and driver support features), defining a transparent process for setting a safety standard, and creating new offences to prevent misleading marketing
- A two-stage approval and authorisation process, developed by building on current international and domestic vehicle approval schemes, and adding a new second stage to authorise vehicles for use as self-driving on UK roads
- A new in-use safety assurance scheme to provide regulatory oversight of automated vehicles throughout their lifetimes, to ensure that they continue to be safe and comply with road usage rules
- New legal roles for users, manufacturers, and service operators, with removal of criminal responsibility for the person in the passenger seat
- Holding manufacturers and service operators criminally responsible for misrepresentation or nondisclosure of safety-relevant information

It should be noted that at the time of writing (September 2022) we do not have a body in the UK specifically to investigate accidents involving autonomous self-driving systems, although the government has stated an intention to include this in a new investigation unit (Gov.uk, 2022):

> *The specialised unit will also provide vital insight into safety trends related to new and evolving technologies, which could include self-driving vehicles, e-scooters and electric vehicles (EVs), to ensure the country maintains some of the highest road safety standards in the world and exciting new technology is deployed safely.*

The European Commission has proposed the AI Liability Directive which would help people harmed by artificial intelligence and digital devices (IET, 2022) by reducing the burden of proof required in litigation.

4 Conclusions

This work has highlighted and recounted some of the historical accidents on the railway in the UK, summarised recommendations made at the time, and listed aspects of those accidents which we regarded as relevant for safety in general and also (more specifically) considered whether they have relevance to the introduction of self-driving vehicles or vehicles with autonomous features on the public road. Perhaps unsurprisingly, we found that we could easily identify numerous causes for these accidents from the past that translate to the introduction of any new transport technology. We acknowledge, of course, that some specific causes are inherently railway focussed (e.g. a boiler explosion), but have nevertheless looked for modern parallels with a potential for danger (e.g. user modifications to technology). We also recognise that many directly relevant causes are covered by modern basic safety practice, but have nevertheless included them because – occasionally – basic safety practice can be forgotten, or even subverted due to business and financial exigencies (Robison, 2021). And we most definitely appreciate that the designers and developers of AV systems may already have identified, and likely addressed, the majority of issues raised here; we nevertheless hope that a reminder of their relevance to the possibilities for catastrophe can, at the very least, do no harm and might even prove helpful.

In section 2 our consideration of the 15 selected railway accidents led us to identify over 75 issues or topics that could be seen as relevant (often directly relevant) to anyone addressing the safety concerns posed by AVs on the road. These were extracted and itemised as the Read Across bullet points after each accident. In this section we will endeavour to present our findings (drawn from this long and, perhaps, daunting list) rather more succinctly by offering a more structured distillation in which we aim to capture the salient elements. The straightforward structure we have adopted is as follows. First, we bring together issues and concerns relating to the environment in which an AV operates; second, we consider the problems that could arise from people engaged with an AV (primarily, but not exclusively, vehicle owners and occupants); and third, we group the dangers and uncertainties inherent in the technology and systems controlling the AV itself. We add a fourth subsection where the focus is on the aftermath of those residual accidents which, we fear, will surely still occur.

4.1 Environmental issues

Designers will always endeavour to take into account the problems created by unusual (special, abnormal, exceptional) external circumstances. Nevertheless, experience indicates that complete coverage is most unlikely, and probably impossible to achieve. Challenging scenarios for AVs will include:

- Modifications to the road infrastructure, such as road works, roadblocks, diversions, damage to the road surface, missing or emergency signage, manual or temporary signalling, special instructions or regulations
- Severe weather, flooding, fire or smoke can clearly impact on visibility and road adhesion
- Unrecognised objects on (or near) the road which will always have the potential to generate confusion for an automated system
- Combinations of the above issues, and/or rapid changes in the scenarios would, of course, add to the difficulty

An obvious option for reducing the impact of these problems would be by providing advance warning to the AV, via a variety of modern communication mechanisms.

4.2 Issues relating to people

Human factors to take into consideration in the design of AV control technology should include:

- The occasions on which AV owners and users will behave unexpectedly – and thus potentially dangerously – because of very many different factors, such as carelessness, complacency, misunderstanding, confusion, stress, panic, foolishness or even malice
- More specific situations that can be anticipated, such as AV occupants disregarding warnings, not following precautionary guidance, leaving the vehicle unattended (when this would be inappropriate), disabling or re-enabling autonomous functions at critical moments, risks arising from misbehaviour of children
- Instruction from authorised persons (e.g. the police in the context of an accident) to contravene the rules of the road (e.g. turn around on a motorway, drive up the hard-shoulder and leave via the entrance)
- Individuals with reduced capability or knowledge, such as the young, the elderly, or those with a physical or mental disability – especially when no fully competent adult is present
- The risks arising from unauthorised attempts to modify AV systems in order to enhance vehicle performance or disable safety features

4.3 Technology and systems issues

This is certainly the broadest area on which commentary can be derived from our set of accidents; furthermore, much of the read across is widely applicable to safety systems in general. For example, indications supporting the need for fail-safe approaches and the provision of "defence in depth" against envisaged hazards. Much of the following is similarly generic, but is emphasised here because of prevalence or significance in the accidents considered above.

- Critical system data must be checked to assure its validity (accuracy and timeliness) – as input from sensors, when maintained and manipulated internally, and as output to actuators
- In similar vein to data checking, ongoing health monitoring of sensors (including recalibration as needed) and built-in-test mechanisms for control systems and vehicle components should be included
- The aim of the above checking and monitoring is twofold: first to detect (and then respond to) any deviation from or breach of the AV's operational capabilities, and second to strive to ensure that the AV's internal model of its environment is an accurate reflection of external physical reality
- Recognition of the multivarious sources of system failure lies at the heart of dependable design practice: communication corruption or disruption, design defects (the bane of software), defective components, natural wear-out, deficiencies introduced via mechanisms added specifically to defend against failure, …
- It seems abundantly clear that the key software component of an AV will be subject to periodic (quite possibly frequent) amendment and upgrading; AV software developers should – and surely will – be fully aware of the risks inherent in incontinent patching of complex systems
- Accidents on the railway have repeatedly shown the necessity for adequate braking systems and this drives us to mention that issue here, to indicate areas where innovative practices may perhaps bring a benefit: response to external alerts, the capability for near-immediate reaction (initially and as the situation develops) – perhaps even the notion that braking could be the default action, applied unless the AV actively and definitively determines otherwise[15]

4.4 After an accident

Given that accidents involving AVs cannot be wholly prevented, capability will be needed for an appropriate response in the aftermath.

- In the event of an accident caused (directly or indirectly) or contributed to by an AV, or inflicted on an AV, or even just nearby, then the first

---

[15] A historical automobile anecdote may be of interest. Ettore Bugatti was upbraided by a client about the inadequate brakes on one of his models. He (allegedly) replied "*I make cars to go. Not to stop.*"

- requirement is for the AV to be aware of the incident, and – as far as possible – identify its characteristics
- The AV should then operate appropriately, to attempt to mitigate consequences, and be available to assist afterwards (for instance, by providing transportation to reach medical help)
- The incident should be notified to nearby vehicles, and to the relevant emergency and police authorities, with a clear indication of severity
- Records of all information (data, messages, events) pertaining to an accident should be retained, and protected from tampering or deletion; ideally these records should be capable of being independently verified
- Following the excellent historical practice on the railway, an independent body should be set up to investigate accidents involving AVs, with responsibility to determine causes, resolve any disputation, and make recommendations; the importance of such a body can hardly be overstated, providing benefits in technological development and in support of litigation – this is much too important to be left solely to developers and manufacturers

There is also the issue of unknowns: many aspects of safely operating AVs are still to be worked out, and the history of the railways clearly shows that some potential hazards will not be obvious until after there has been an accident as a result. What is also not at all clear at this juncture is how conventional drivers will react to AVs that they encounter: in most cases they will surely be indifferent, but it seems quite plausible that a cavalier minority of manual drivers will behave badly (for example, by posing a "challenge" to an AV's braking ability) and thereby generate dangerous situations.

We therefore conclude that we can anticipate that many accidents will occur with AV technology on the roads, but we can hope that some of these might be averted by a careful consideration of historical precedents.

5 Further Work

Further work could address other significant transport accidents from the past. We have covered only a small number of serious rail accidents in the UK; unfortunately there is no lack of other candidates that could be useful to consider. And then, of course, accidents are not confined to just railways in the UK – looking at an international dimension could be worthwhile, or even a parallel consideration of maritime and aviation accidents.

One area that could be fruitful is the analysis of security aspects on the railway, and how these have changed over time. In particular, the impact of basic security measures (e.g. trackside fencing, CCTV) on safety and accident reduction could be analysed. Cybersecurity issues are now a vital concern in all domains; it would be valuable to look at how techniques adopted, or under development, for other transport sectors could augment solutions being incorporated in AVs.

Lastly, as we noted in section 3, there is a major opportunity for making appropriate connections with work on the creation and evolution of legislation and regulations in the rail sector, and its influence and impact on safe working practice.

Disclaimer Only one author can claim any specialist expertise in AVs, and all lack professional knowledge of railway history or railway practice (but will admit to being railway enthusiasts). All opinions are those of the authors and not their respective organisations.

References

Autonomous road vehicles (2022), Guidance and Research,
 https://www.gov.uk/transport/autonomous-road-vehicles , accessed October 2022
Earnshaw, Alan (1996), Trains in Trouble, Atlantic Transport
Gingell, Carol (2014), Thorpe St. Andrew Rail Disaster 1874, Broadland Memories web site,
 https://www.broadlandmemories.co.uk/thorperail1874.html , accessed October 2022
Gov.uk (2022), Government launches country's first ever investigation branch focused on road safety, https://www.gov.uk/government/news/government-launches-countrys-first-ever-investigation-branch-focused-on-road-safety , accessed October 2022
Hall, Stanley (2006), Railway Milestones and Millstones: Triumphs and Disasters in British Railway History, Stanley Hall, Ian Allen Publishing
IET (2022), mailing, EU to facilitate suing AI and drone companies in new draft rules, https://eandt.theiet.org/content/articles/2022/09/eu-to-facilitate-suing-ai-and-drone-companies-in-new-draft-rules/ , accessed September 2022
ISO/FDIS (2022), Road vehicles – software update engineering,
 https://www.iso.org/standard/77796.html , accessed October 2022
Lawcom.gov.uk (2022), Key Recommendations, https://www.lawcom.gov.uk/project/automated-vehicles/ , accessed September 2022
Morse, Greg (ed.) (2020), Leading Health and Safety on Britain's Railway,
 https://www.rssb.co.uk/en/safety-and-health/leading-health-and-safety-on-britains-railway , accessed October 2022
NetworkRail (2022), Lessons Learnt, https://safety.networkrail.co.uk/tools-resources/safety-bulletins/ , accessed October 2022
Nock, O.S. (1970), Historic Railway Disasters, Arrow Books
Prosser, Ian and Keay, David (2019), A New Illustrated History of Her Majesty's Railway Inspectorate from 1840, Steam World
Railways Archive (2022a), Railways Archive web site, https://www.railwaysarchive.co.uk/ , accessed August 2022
Railways Archive (2022b), 1849 Plympton accident,
 https://www.railwaysarchive.co.uk/docsummary.php?docID=1562 , accessed September 2022
Railways Archive (2022c), 1860 Tottenham accident,
 https://www.railwaysarchive.co.uk/docsummary.php?docID=4636 , accessed August 2022
Railways Archive (2022d), Report of the Court of Inquiry into the Collision near Norwich on 10th September 1874, Tyler HW, HMSO

https://www.railwaysarchive.co.uk/documents/BoT_Norwich1874.pdf , accessed October 2022

Railways Archive (2022e), 1909 Cardiff accident, https://www.railwaysarchive.co.uk/docsummary.php?docID=318 , accessed September 2022

Railways Archive (2022f), Further Reading, https://www.railwaysarchive.co.uk/books.php , accessed October 2022

Rainhilltrials (2022), https://rainhilltrials.co.uk/william-huskisson-and-the-first-railway-tragedy/ , accessed August 2022

Robison, P. (2021), Flying Blind: The 737 MAX Tragedy and the Fall of Boeing, Penguin Business

Rolt, L.T.C. and Kichenside, G.M. (1986), Red for Danger, Pan Books

Wikipedia (2022a), 1857 Lewisham rail crash, https://en.wikipedia.org/wiki/1857_Lewisham_rail_crash , accessed August 2022

Wikipedia (2022b), 1861 Clayton Tunnel rail crash, https://en.wikipedia.org/wiki/Clayton_Tunnel_rail_crash , accessed September 2022

Wikipedia (2022c), 1862 Winchburgh rail crash, https://en.wikipedia.org/wiki/Winchburgh_rail_crash , accessed September 2022

Wikipedia (2022d), 1876 Abbots Ripton rail accident, https://en.wikipedia.org/wiki/Abbots_Ripton_rail_accident , accessed September 2022

Wikipedia (2022e), 1889 Armagh rail disaster, https://en.wikipedia.org/wiki/Armagh_rail_disaster , accessed September 2022

Wikipedia (2022f), 1898 Wellingborough derailment, https://en.wikipedia.org/wiki/Wellingborough_rail_accident , accessed August 2022

Wikipedia (2022g), 1910 Hawes junction rail crash, https://en.wikipedia.org/wiki/Hawes_Junction_rail_crash , accessed August 2022

Wikipedia (2022h), 1915 Quintinshill rail disaster, https://en.wikipedia.org/wiki/Quintinshill_rail_disaster , accessed August 2022

Wikipedia (2022i), 1921 Abermule train collision, https://en.wikipedia.org/wiki/Abermule_train_collision , accessed August 2022

Wikipedia (2022j), 1940 Norton Fitzwarren rail crash, https://en.wikipedia.org/wiki/1940_Norton_Fitzwarren_rail_crash , accessed August 2022

Wikipedia (2022k), Mont Blanc tunnel fire, https://en.wikipedia.org/wiki/Mont_Blanc_tunnel_fire , accessed October 2022

Making the safety standards work for you

Stephen Bull, Lucia Capogna, Rob Davies, Stephen Gill

Ebeni Limited, Corsham UK[1]

Abstract *Every safety-critical industry has its own legislation and standards which safety engineers have to consider in their demonstration that a system is safe. Often these standards overlap and conflict and they inevitably become out of date as technology and understanding advance. In addition, there are always trade-offs to consider between safety, environment, performance (and sometimes security). Choosing an approach for a project involves selecting standards and resolving these conflicts in a way which is acceptable to each of the parties whose approval the project needs to secure. This paper examines some of the challenges, drawing on the authors' experience across multiple industry sectors. The paper then derives some principles for making suitable, informed choices at the outset of a project and for monitoring those choices as the project proceeds.*

1 Introduction

1.1 Why do we need standards?

Standards are an essential means to establish good practice for system development and for the demonstration of safety. They can reduce repeated effort (and therefore reduce costs) by establishing common approaches, especially for devel-

---

[1] The authors are engineers with experience of safety and cybersecurity engineering and assurance in a variety of safety-critical industry sectors, including rail, nuclear and aviation (civil and military). All authors can be contacted by email in the form forename.surname@ebeni.com

© Ebeni Limited 2023.
Published by the Safety-Critical Systems Club. All Rights Reserved

opment of products and systems which are similar to those which have been developed previously. Standards can also bring safety benefits, by capturing lessons learned from accidents and incidents. (E.g. the introduction of TPWS[1] on UK railways to protect against train collisions.)

Safety legislation defines the criteria by which a development is judged to be safe. Legislation recognises the role of standards by acknowledging that application of an established standard is a valid way to develop a product or system which will meet the safety criteria.

There are two principal ways where this approach (application of standards to achieve safety) can be insufficient:

- where the state of the art has improved beyond the requirements of the standard, such that developing to the standard does not achieve the best level of safety which can be reasonably achieved
- where suitable standards have not yet been developed, either because the application is novel within an established sector, or because the sector itself has not yet established system safety standards

It is easy for system developers to become complacent and attempt to apply standards as the "default approach", assuming that "compliant" is equivalent to "safe". The defence against this is the requirement to undertake explicit safety assessment of any new development or change. In addition, regulators require various forms of monitoring to examine whether better levels of safety can be achieved and to direct further developments accordingly. In UK rail, this includes the work of the Rail Safety and Standards Board (RSSB) as well as the less well known "CSM for Monitoring". In the UK nuclear industry, licensed installations are required to undertake periodic safety reviews.

1.2 The problem of choosing the right standard(s)

The fundamental requirement is to comply with the law: in the UK[2], the key primary legislation is the Health and Safety at Work Act 1974 [1], which requires that risk is reduced to a level which is As Low As Reasonably Practicable (ALARP)[3]. This is supported by further regulations (which must also be complied

---

[1] Train Protection and Warning System – an automated system to apply brakes if a train is overspeeding or passes a red signal: introduction was legislated following inquiries into major UK railway accidents.

[2] Although the specific legislation varies between the countries of the UK, the approach is the same.

[3] This is the concept of weighing the level of safety risk against the trouble, time and money needed to control it, in order to determine a level which is acceptable and justifiable [2]. The Act uses the term "so far as is reasonably practicable", but the result is the same.

Making the safety standards work for you 59

with) and standards (which define approaches for achieving safety for particular scenarios or types of development). The challenge is to identify and apply a suitable, consistent set of standards which supports the project in achieving the overall target.

The task of choosing an appropriate set of standards is made more difficult by several factors, including:

- inconsistency or conflict between standards (and regulations),
- differing approaches to demonstrating safety, especially where a project intends to apply products developed elsewhere,
- projects spanning multiple disciplines or application areas.

Choice is often constrained by the client, by requiring specific standards to be applied.

During the lifetime of a project, the choices made may be further challenged by:

- changes to standards,
- advances in technology or practice which may go beyond existing standards.

All these issues need to be successfully navigated in order to demonstrate that a given development achieves the overarching safety requirement.

1.3 Purpose and scope of this paper

In this paper, we elaborate on some of the issues outlined above, drawing on the authors' experience; we also provide some practical guidance on addressing these issues in order to successfully demonstrate that a project achieves the required safety targets.

The authors' main experience is in application of standards to demonstrate the overarching ALARP requirement in UK legislation in railway, cyber and nuclear sectors. Although the detailed arrangements in other countries are different, we believe that safety practitioners face similar problems.

The focus in this paper is on the process of demonstrating the functional safety of engineering systems, and on the application of relevant standards to do this.

1.4 Standards and legislation

In the UK the main overarching legislation is the Health and Safety at Work Act 1974 [1] as mentioned above, supported by the Managing Health and Safety Regulations 1999 [3].

Other legislation[4] gives guidance to particular industry sectors, and there is sometimes a grey area where legislation is treated as if it is a standard, defining processes for achieving safety.

Examples of key (UK) legislation in this area include:

- CSM-RA[5] [1] defines the approach which must be applied to evaluate the safety of any "technical, operational or organisational" change to any mainline railway within an EU member state[6].
- ROGS[7] [5] are the UK regulations for railway safety, which apply to all railway duty holders, and establish requirements for safety management systems and safety assessment.
- CDM[8] [6] are the UK regulations which manage health, safety and welfare relating to construction projects (including new build, demolition, refurbishment, repair and maintenance).
- Nuclear Installations Act [8] defines the licensing of all non-military UK nuclear sites and includes provisions (in part 7) for ensuring safety of such sites.
- Commission Implementing Regulation 2017/373 [7] lays down common requirements for the provision of air traffic management and air navigation services and for the competent authorities which perform certification, oversight and enforcement tasks in relation to these services, and the rules and procedures for the design of airspace structures.

Standards define established ways for achieving particular goals, including development of safe systems. Whereas legislation is mandatory, the application of standards is open for interpretation and negotiation, dependent on the requirements of (all) stakeholders. Key standards considered in this paper include:

- EN 50126 [9] for assuring reliability, availability, maintainability and safety in the development of railway technical systems

---

[4] The term legislation includes both Acts of Parliament (primary legislation) and Regulations (secondary legislation, enacted under powers given by primary legislation).
[5] Common Safety Method for Risk Evaluation and Assessment
[6] Imported into UK law under the Brexit regulations
[7] Railways and Other Guided Transport Systems (Safety) Regulations 2006
[8] The Construction (Design and Management) Regulations 2015

- EN 50128 [10] / EN 50657 [12] for development of safety critical software for railway systems[9]
- EN 50129 [11] defines a format for safety cases for railway systems and provides guidance on architectures and development processes
- IEC 61508 [13] defines series of sector-independent standards specific for functional safety of Electrical/Electronic/Programmable Electronic Safety-related Systems: it comprises 7 parts that defines methods for designing, developing, deploying and maintaining such safety-related systems
- ISO 27001 [14] for security of information technology systems, including requirements at an organisational level
- IS1&2 [30] providing UK guidance for compliance with ISO 27001, although no longer maintained, they are retained as legacy documents
- Safety Assessment Principles for Nuclear Facilities [28] providing the principles by which nuclear safety should be assessed at UK licensed sites

A longer description of the legislation and standards listed above is given in Appendix A.

2 Problems with standards

The following sections present a number of issues experienced with applying safety standards and legislation on development and change of systems relating to functional safety. The specific examples are drawn from the experience of the authors, but we believe the issues are equally applicable across other safety critical sectors.

2.1 Safety silos

Safety is an emergent property of a system. Where standards or legislation are aligned to a particular discipline or area of the system, they can only influence safety for that part of the system. Full treatment of safety needs to take a truly *system level* viewpoint to demonstrate that the overall risk is reduced to an acceptable level.

This problem can be illustrated by some examples:

---

[9] Note: these standards are very similar, although the differences do have an effect on the processes which need to be followed: they apply to signalling and rolling stock systems respectively. They are in the process of being updated into a single standard which covers both.

1. A main safety function of a railway signalling system is to prevent collisions between trains – such systems are usually developed to CENELEC[10] standards (EN 50126 / 8 / 9) and are usually "fail-safe". If a signalling system fails, the "fail safe" principle means that trains are signalled by alternative means (or stopped entirely), significantly reducing the capacity of the service, leaving many people to travel by alternative means. Temporary bus services may be provided, leading to risks where drivers are brought in at short notice to drive routes with which they are not familiar. The failure can also have a knock-on effect on risk at stations, due to overcrowding and passengers rushing to catch trains. Thus a signalling system failure may lead to an *increased* risk to passengers, which cannot be managed simply by better application of the CENELEC standards.

2. In the nuclear power industry, there are multiple stakeholders competing to optimise 'safety'. The focus is often on balancing nuclear safety with environmental safety, requiring discussions between the nuclear regulator and the local environment agency (see guidance on UK Government website [15]). However, conventional health and safety (of workers at the nuclear plant – governed in the UK by the CDM regulations) and security also need to be considered. Each set of regulations only governs part of the risk: demonstration that the overall risk is acceptable requires an overall system level approach, and is likely to require compromises in one or more of the governing regulations. Construction and operation of a new nuclear power station in the UK requires site specific permissions from the UK Office for Nuclear Regulation (ONR) and environmental permits from the UK Environment Agency (EA). For example, there is a requirement for nuclear sites to have backup power supplies to ensure nuclear safety in the event of electrical supply failure. However, where this backup is provided by diesel generators, there are environmental requirements on diesel storage and exhaust emissions. These requirements are in tension because the provision and regular testing of diesel generators increases the environmental risk while at the same time managing the nuclear safety risk.

Our tendency to work in silos exacerbates the problem, because conflicts do not become apparent until late in the development when the project is so far advanced that major changes are prohibitively expensive.

There is no magic bullet, and the best way to address these conflicts is to recognise the issues at the start of the development and establish regular discussions between stakeholders (including regulators) to find mutually acceptable solutions. Each discipline should be aware of the goals of others and identify issues

---

[10] European Committee for Electrotechnical Standardization

which may affect others. These issues should be communicated between disciplines on a regular basis, to ensure that they are properly considered in a timely manner. Examples include:

- cyber security analysis should be aware of the goals of system safety and report any threats identified which could lead to harm to people, even if these threats are not analysed in detail by the security team,
- those analysing construction safety issues (under the CDM regulations) need to identify any risks which will not be closed out during the construction process and communicate them to the system safety team in good time to allow operational mitigations to be identified and agreed.

We recognise this is not new advice – but it is still unusual for projects to successfully establish such arrangements early in the lifecycle and follow them through.

2.2 System development vs operational change

CSM-RA is the legislation applicable whenever you make a change to the railway[11]; the CENELEC Standards EN 50126 / 8 / 9 apply to the development of technical systems.

One of the authors has worked on several projects where an early decision had been made to adopt CSM-RA as the overriding approach on the basis that the project was a railway project. Closer examination showed that the project did not (itself) represent a change within the scope of the regulation. (I.e. the project was not (directly) making a change to the railway.)

These projects were developments of technical specifications or support systems which would then subsequently be used as part of a change to the railway. Adoption of CSM-RA in these cases led to a focus on the requirements of that legislation, rather than focusing on the more technical requirements of the CENELEC standards which, in retrospect, would have provided better guidance on how the systems should be developed. In these cases it may have been better to adopt the CENELEC standards directly, along with their guidance both on development processes and design principles (neither is provided by CSM-RA).

The eventual application of these technical developments still needs to apply CSM-RA at the point where an actual change to the railway is proposed and implemented. This application of CSM-RA is best focused on the application of the products into a wider overall railway system, which will (almost certainly) also include changes to processes and organisation. It is the authors' view that, as long as the technical development has applied the CENELEC standards (including the

---

[11] The legislation explicitly includes technical, organisational and operational changes in its scope.

requirement for independent assessment), then sufficient evidence (for the technical development) will be available to support the safety analysis under CSM-RA.

One key challenge is how to structure the safety analysis of the overall change. The CENELEC standards give detailed advice on development of technical systems, including definition of a system lifecycle, guidance on the safety analysis of technical changes and guidance on technical design decisions for safety critical systems. In contrast, CSM-RA provides only a "framework" for safety analysis, giving no detailed guidance. This is understandable, because of the broad scope of changes covered by CSM-RA, but it does mean that the safety analyst must look elsewhere to determine the most appropriate methods to undertake safety analysis of the overall change. The methods applied need to take into account all aspects of the overall change, including changes to process, human roles and organisation, as well as the changes to the technical system.

These aspects of a change are not covered by standards in the same way that technical changes are covered by the CENELEC standards. The analyst has to select from available guidance, dependent on the specific features of the system. This guidance includes:

- Systems-Theoretic Process Analysis (STPA) [20], which is a technique looking at unsafe interactions between system components, moving away from the assumption that an element of the system has failed
- BowTie modelling, which has successfully been used to model and illustrate hazards, causes and consequences, especially in operational risk scenarios (see for example RSSB[12] BowTie Hub [17])
- Guidance from the SCSC Service Assurance Working Group [21], which makes some high-level suggestions for analysis techniques, although the details of the techniques need further development.

These sources provide some assistance to the analyst, but there is also space for more rigorous guidance in this area.

Both CSM-RA and CENELEC require an independent assessment to be conducted to confirm that the requirements have been correctly applied. The approach taken by the assessor significantly affects the value gained from such an assessment. Ideally, the assessor will provide an independent professional opinion of whether the underlying goal (of demonstrating the safety of the system) has been achieved, but it is very easy to fall back on a "tick box" approach, where the assessment activities are reduced to checking for evidence that each clause of the regulation or standard has been applied. (In fact, one of the authors is involved in a contract where the assigned task is specifically to confirm whether the development conforms to the relevant standard, rather than expressing an opinion

---

[12] Rail Safety and Standards Board – the organisation charged with helping the British rail industry to work together to drive improvements in the railway in Great Britain

on the demonstration of safety.) The most benefit will be gained for the project if the assessor takes a constructive and pragmatic approach to demonstrating safety (rather than demonstrating all the boxes are "ticked") and has sufficient experience in the domain: (a) to focus on the key issues and (b) to form a judgement over whether they have been resolved effectively.

These experiences show that it is important to choose the appropriate standards and guidance for the level and type of change involved, partly depending on whether it is a change to a technical product or a socio-technical system.

2.3 Cross acceptance

Each sector has its safety standards, and a product developed for a particular industry will usually be certified to that industry's standards and accepted under the regulations in a certain country. However, as companies seek to diversify into multiple industries and across global markets, they want to sell their products as widely as possible. But demonstrating compliance to additional standards after a product has been developed is often a challenge, because the precise requirements of the standards or legislation are different (or differently interpreted), and because the requirements often relate to the processes by which the product is developed.

One of the authors was involved in a railway project which utilised products originally developed to non-railway standards. One example was a programmable logic system developed to machinery control systems standards (EN 13849-1 [22] and EN 62061 [23]). The project requirements included development to the railway-specific EN 50126 / 8 / 9 standards. In this case, it was not feasible to directly demonstrate compliance with the railway standards, so the author created an argument that an equivalent level of safety was achieved, based on the application of good safety engineering practice to evaluate the safety functions delivered by the logic system and to ensure that the safety risks were suitably mitigated. The argument was made easier by the fact that the standards involved are derivatives of IEC 61508 and therefore employ similar basic concepts. (However, it should be noted that EN 50128 contains more detailed activities and requirements to fulfil specific railway system needs; thus compliance with IEC 61508 is not automatic and gap analysis is required.) The most difficult and time-consuming part of the process was actually convincing the client chief engineer (whose specialism was railway systems, rather than safety of programmable systems) that the argument was suitable.

Another of the authors faced an opposite challenge: software developed and assessed to EN 50128 was required to show IEC 61508 compliance. Gap analysis was still required, but it was found that no further assessment was needed because the software already met the IEC 61508 requirements.

These approaches are examples of cross-acceptance, which is based on the assumption that if a product has already been accepted for use, then application in a new environment should only need to consider the differences between those environments and their effects on the operation (and safety) of the product. (Environment here needs to be interpreted very widely, including operational and maintenance practices, as well as the underlying assumptions within the standards used to develop the product.) The principles and process of cross-acceptance are captured well in a (now withdrawn) CENELEC technical report [24] which was originally published in 2007 as a companion to EN 50129: the principles are:

1. Establish a credible case for the baseline application.
2. Specify the target environment and application.
3. Identify the key differences between the target and native cases.
4. Assess the risks arising from the differences.
5. Produce a credible case that the adaptations adequately control the risks.

Although this seems like a simple concept, making a successful cross-acceptance argument can be fraught with difficulty. Often specifying and analysing the differences between environments is a much more difficult process than it might appear. For further discussion see Rod Muttram's paper [25] for the IRSE[13]: he concludes that cross acceptance can be a useful approach which may save acceptance effort, but that the effort required to make the cross acceptance argument should not be underestimated; this is especially true where the underlying principles of the standards differ significantly (as is often the case between the US and Europe). This aligns to the experience of one of the authors who has been involved (in different roles) in several attempts at cross acceptance of railway signalling equipment.

2.4 Common approach

Organisations often need to work under multiple sets of standards or regulations. This can be because multiple standards or regulations apply to a particular development (e.g. both CDM and CSM-RA) or because different activities which the organisation undertakes fall under different regulations. For example, Transport for London (TfL) operates trains on multiple railways: only some of these are in the scope of CSM-RA. One implication which this has is in the arrangements for independent assessment:

- changes in the scope of CSM-RA require a significance test, and "significant" changes must be assessed by an Assessment Body (AsBo),

---

[13] Institution of Railway Signal Engineers

- changes outside the scope of CSM-RA come under the overarching ROGS regulations, which require a different test for significance, where "significant" changes must be assessed by an Independent Competent Person (ICP).

The significance tests and assessors play similar roles but the details of their application are subtly different.

For reasons of efficiency, an organisation like TfL may wish to apply a single approach to demonstration of safety, while providing variant paths through the process to cater for the different regulations or legislative frameworks. This approach needs to be embedded in the organisation's Safety Management System (SMS) and approved by the relevant regulators. The approach should also be supported by detailed guidance which explains how and when it should be applied, including clear explanation of when the different variants on the processes are applicable. It is also important that the SMS is fully integrated with the organisation's wider management systems, to ensure that the decisions relating to safety are fully aligned to and supported by other processes, such as the engineering lifecycle and the artefacts produced by it.

On several projects, we have found that CSM-RA forms a good framework for safety management on rail projects. The requirement for rigorous hazard analysis and the risk acceptance principles within the legislation are well aligned with other approaches (e.g. TfL's S1521 [26] and EN 50126). As discussed earlier, the biggest problems come where CSM-RA is treated as a detailed standard, rather than a framework. In order to define a successful safety management process, CSM-RA needs to be supplemented by detailed guidance appropriate to the type of change being made: for a technical change this could be EN 50126 and associated standards.

Another example comes from the nuclear sector, and the search for a suitable approach for new nuclear submarine design. The UK has a strong pedigree in the design, manufacture, operation of nuclear submarines and iterative improvement of designs brings technical performance benefits, increased capability and improved safety. The recent revitalisation of civil nuclear in the UK has raised the question of what the baseline should be for new nuclear submarine design:

- Should this be based on civil nuclear best practice, modified as appropriate to apply to a submarine and warship?
- Or should the approach build on the pedigree of past military experience and benchmark against civil?

This has caused much debate in the development of regulation and the definition of 'best practice' or 'Relevant Good Practice' for new submarine design and build, covering topics such as:

- What is best (or good) practice, and how is it captured in the standards?

- How should the engineering community work together to agree a suitable set of standards?
- How should we cater for varying interpretations of the standards?
- How should we assure compliance with the agreed standards?
- How do we evaluate and capture evolving best practice?
- How do we address individual bias arising from past experience?

Overall, a common approach across an organisation can bring benefits. These include increasing understanding and resource flexibility. But care is needed to ensure that the approach is:

- simple enough to be easily applied,
- flexible enough to work under all the legislative regimes,
- updated as required when practices or legislation change.

2.5 Standards Obsolescence

Another issue to consider when selecting standards is obsolescence. A standard is based on the best knowledge and practice of the authors at the time that it is written. As time passes, knowledge advances and new methods are developed. Hence, as a standard gets older, it (often) becomes less representative of best practice within the industry. This is especially true in fast-moving disciplines such as electronic technology. (As an example, the latest update [16] to EN 50128/ EN50657 will replace the current list of programming languages with a list of principles, in an attempt to reduce future obsolescence of the standard.)

There are plenty of standards still in regular use which have not been updated for many years; examples include:

- ARP4761 [18] (Guidelines and Methods for Conducting the Safety Assessment Process on Civil Airborne Systems and Equipment), not updated since originally published in 1996,
- ED-80 [19] (Design Assurance Guidance for Airborne Electronic Hardware), also still in its original version, published in 2000.

To counter this obsolescence, standards are reviewed and updated if required. In some cases, particularly in rapidly developing fields like cyber security, there can be a large number of changes between revisions of the standard. For example, the set of cyber security requirements in ISO 27001 [14] was originally published in 2005 and underwent a major revision (published 2013) with a large number of the requirements being either modified, deleted or added to. (A subsequent revision was published in 2017.) Similarly, IEC 61508 (latest edition published 2010) is in the process of being updated and the new edition is expected to address more recent developments such as multicore processors, which are now in widespread use. The magnitude of such changes means that adopting the new version of a standard can involve significant costs for a manufacturer, meaning that they will often resist adoption for as long as possible. Even this review and update process is not straightforward, as standards refer to each other and are often not updated in parallel: this leads to (sometimes significant) inconsistencies.

Whilst it may seem an obvious decision to move to a new standard once it is revised, there are a number of reasons why this may not be practical. For example, the standard may have been published later in the project lifecycle, where implementing the new standard would incur expensive redesigns. One author is working on a project where, while the project was in the manufacture stage around ten years after its initiation, there was a major revision of the client's own internal security standard for how to achieve security accreditation. Implementing this new version would have caused vast disruption to the project, including an expensive change request to the supplier (the previous version was written into their contract): firstly, to undergo an assessment to determine whether design changes were required and then, if necessary, to implement the changes. As a result, gaining endorsement from the accreditor to stick with the older version allowed the project to stay on track and not incur any additional expenditure.

Whilst this did allow the project to progress on time, some terminology issues were encountered when implementing the previous standard which required use of the 2005 version of the ISO 27001 [14] baseline control set. As this version of the standard was based around the IT terminology of that time (seventeen years ago) the author found some disconnects between the standard and what is used now. For example, the control set has requirements for teleworking (out of office but in a similar area) but none for remote working (out of office in any area even globally) as this was not common practice at the time[14]. The author solved this issue by gaining endorsement that this control should be interpreted to cover all remote working and compliance statements generated to cover this.

Another example can be seen in the ONR Safety Assessment Principles (SAPs) [28], which distinguish approaches between new facilities and facilities built to earlier standards. The principle (no. 33) relating to facilities built to earlier standards includes: "The extent to which the principles ought to be satisfied must also take into account the age of the facility or plant. For facilities designed and constructed to earlier standards, the issue of whether suitable and sufficient measures are available to satisfy ALARP will need to be judged case by case."

---

[14] It should be noted that even in the 2017 version of ISO 27001 [14] there is still reference to teleworking rather than remote working.

2.6 Standards Validity

Sometimes a project attempts to head off standards obsolescence by considering standards still in preparation: this approach also has drawbacks.

One of the authors has been involved in an independent software assessment against the proposed new part 5 of EN 50126 for software [27]. When the contract between the customer and the manufacturer was established, this standard was very mature but still in preparation. Both parties agreed on the selected document, however it was never accepted by the national committees and therefore has never become a formally published standard.

The project continued to apply the draft standard as it was requested by the contract. The main issue is that the software functions implemented will be assessed against a standard that is not recognised as valid. Additional effort (and cost) would be required (as outlined in the section above on cross-acceptance) to demonstrate compliance to current standards.

2.7 Standards Accessibility

Another barrier to application of standards is cost, especially in an environment where many different organisations (often including multiple small consultancies) are involved in a project. Although some standards are freely available, either to all (e.g. UK Defence Standards) or to registered suppliers (e.g. Network Rail Standards) there is increasing reliance on standards developed by international standards bodies: these are usually expensive to obtain and this is made worse when they are published in multiple parts. To illustrate this, the authors looked into what it would cost to purchase all possible parts of some common sets of standards at face value from the publishers' webstores. This is shown in Table 1.

Table 1. Standards Pricing (autumn 2022)

Standard Series	Number of Parts	Cost
IEC 61508	7	3285 CHF (£2957)
IEC 62443	9	3085 CHF (£2777)
ISO 27000	58	8024 CHF (£7222)

For large organisations this is often managed by comprehensive corporate standards subscriptions. However, for smaller organisations this cost can be prohibitive and potentially risky if it is unknown how useful the standard would actually be. For example, within the ISO 27000 series, ISO 27001, 27002 and 27005 are well known and widely used. However, the other parts are less well known and, without reading them, it is difficult to judge how applicable they would be to a user's direct need.

An additional constraint on the use of standards is the implementation of digital rights management on downloaded copies which limits who can view them and whether they can be printed or not. This results in either having to purchase the standards for every individual employee who requires them or entering into a multi-user subscription arrangement. The situation is further exacerbated by practices where organisations licensing standards impose conditions on who (within the receiving organisation) is permitted to use the licence: in one example, a licence is only available to direct employees of the receiving organisation, such that staff working as consultants do not have access to the standards they need to complete their assigned tasks. As a result, old versions of standards are often applied if an organisation is either not willing or unable to meet the licensing terms of the new standard.

An example of this has been the continued application within the defence sector of the UK Government Information Assurance Standards No 1 and 2 (IS1&2) [30] and accompanying supplement. This standard was last updated in 2012 and is no longer maintained by its author (CESG[15] – the former name of NCSC[16]) who has archived it as a legacy document. As the baseline control set detailed within the IS1&2 supplement is based on ISO 27001:2005, the base requirements are now seventeen years old. One of the reasons for the continual use of this document within the defence sector is that, as it was developed by the UK government, there are no licence fees attached to it and it can be distributed at will to all client and supplier organisations. Although ISO 27001 underwent a major revision in 2013, this version and subsequent later ones are not free to obtain or distribute. Hence, the older and free IS1&2 has continued to be used. In this instance, recently there has been a light at the end of the tunnel with a directive to move away from IS1&2 and ISO 27001 and towards the National Institute of Standards and Technology (NIST) set of cyber security standards (e.g. NIST SP 800-53 [31]) which are, like IS1&2, free to access for all.

3 How to make the standards work for you

In the previous section we identified some common issues faced in the application of standards when developing and deploying safety related systems. We believe that the best approach to addressing these problems is to start early and

---

[15] Communications-Electronics Security Group, a group within the UK Government Communications Headquarters (GCHQ)

[16] National Cyber Security Centre

consult widely, reaching an agreement with the key stakeholders (including engineering teams, clients and regulators). The project's approach should be documented in a Safety Justification Strategy or similar which:

- **defines the overall safety target(s)** which the project must achieve: these may be derived from generic national guidance, or the project may have specific targets based on the organisation's own SMS, or on specific safety improvements which it needs to achieve,
- **defines an underlying approach** for how safety will be managed: this could be as simple as identifying the key stages of safety analysis and management: system definition, hazard identification, hazard analysis, definition of requirements, safety demonstration and acceptance,
- **identifies the regulations and / or Safety Management System** under which the project will be accepted and who is responsible for granting acceptance,
- **identifies the standards to be applied**: this should be specific about versions of standards, the areas of the project to which each standard will apply and whether the standard will be applied in whole or in part (especially where a standard is being applied outside its usual / intended area of application); where incorporating products developed to other standards, the feasibility of cross-acceptance should be established,
- **identifies conflicts** between regulations or standards and how they will be resolved,
- defines **how the project will deal with changes** to legislation or standards which occur during the course of the project,
- defines a **communication / consultation plan** to ensure that alignment is maintained with all stakeholders throughout the project,
- defines **how and when safety will be demonstrated**: what form of justification will be produced (e.g. CENELEC specific application safety case) and at what stages in the project (usually at each point where there is a significant change in the safety risk),
- defines **what independent review will be undertaken** to support acceptance: in many cases this is prescribed in the regulations, but the precise scope of the review should be defined; when there are multiple independent reviewers (e.g. ISA and AsBo) their interaction also needs to be defined.

The strategy needs to be owned at a high level within the project (usually the system integrator) and accepted by all stakeholders. In the authors' experience (when working for subsystem suppliers), the system integrator may not initially understand their crucial role in defining safety strategy and achieving safety acceptance: it is worth the effort to establish this understanding at an early stage.

The strategy needs to be grounded in a firm understanding of what the project aims to achieve and (eventually) the solution which will be adopted (including both technical and operational aspects). This understanding will grow as the project proceeds through its early stages: as a result, development of the Safety Justification Strategy will be iterative during these stages.

The Safety Justification Strategy will only be successful if it is integrated with the rest of the system engineering activities, as trade-offs will inevitably be required between safety goals and other objectives, such as reliability and security. In addition, effective integration into the requirements management and V&V processes can significantly reduce the costs of demonstrating that safety requirements are implemented, and this is key evidence to support the safety case.

The Safety Justification Strategy needs periodic review throughout the programme to evaluate (a) the effectiveness of the approach; (b) the impact of any changes to standards. Note: publication of new standards does not necessarily mean that the project's approach or standards baseline needs to change: but the impact of the new or revised standards does need to be evaluated and the project's approach agreed with the relevant stakeholders.

4 How standards can be improved

"Design by committee" is a pejorative term for a project that has many designers involved but no unifying plan or vision. Wikipedia entry [29]

A criticism often levelled at standards is that they are "designed by committee". Members of standards committees have their own individual views and motivations, and it can be difficult to reach a consensus on the content of a standard. Representation on committees is often heavily biased towards large manufacturers, who have a vested interest in making requirements less onerous or allowing multiple alternatives in order to align to manufacturers' existing approaches. Standards may benefit from a shift of focus towards a pragmatic approach focused on the aim of the standard: better representation from regulators or independent experts may assist in this shift of focus.

Another concern is the development of overlapping or inconsistent approaches: for example EN 50128 and EN 50657 (relating to the development of safety related software for railway applications, respectively, in signalling and rolling stock domains) are developed and maintained by separate standards sub-committees although they have very similar content. (Note: these standards are in the process of being merged, and the new standard [16] will replace both of these existing standards.) A move towards more commonalities would reduce the cost of standards maintenance and make it easier to develop products which span more than one application area.

A similar point can be made about co-ordinating the development of standards. Examples include the development of:

- EN 50126 / 8 / 9, the current versions of which derive from an earlier version of IEC 61508 and are therefore inconsistent with the current version;
- IEC 62443 [32], which has multiple parts published over at least 11 years, leading to inevitable inconsistencies, not least because of the advance in technology over that period.

The overall process of applying standards would be simpler if these updates could be co-ordinated to give a consistent set of guidance. The authors recognise that this is not always practicable, but it should be included as an aim in any standards update.

As a final plea in this section, a way is needed to make standards more accessible, so that practitioners can access the standards applicable to their work. The authors recognise that there is a cost involved in developing standards (although the experts are not paid by the standards organisations for their time), but it is not acceptable when this becomes a price / licensing barrier for those who need access.

5 Conclusions

In this paper we have highlighted that standards promote the documentation of good practice, leading to reduced costs by establishing common approaches; they can also bring safety benefits, capturing lessons learned from accidents. However, because human understanding and technology are constantly advancing, standards are always out of date to some extent. In addition, system development always has conflicting goals, especially between safety, environment and performance (and sometimes security) – this conflict is also reflected in the standards developed for these disciplines.

We have highlighted some of the challenges with the application of standards, including:

- choice of standards which align to the level and type of development being undertaken,
- conflict between different parts of a system and different disciplines,
- the relative lack of detailed standards, especially in the operational and process aspects of a change,
- application of products developed in a different industry area under different standards,
- standards obsolescence and update,
- inconsistency and conflict between standards themselves.

Drawing these issues together, we believe that it is critical to develop a safety justification strategy which sets out how a project will achieve its overarching safety goals, the standards it will apply, and how it will manage conflicts and updates. Acceptance of this strategy should be obtained from all stakeholders, including the regulators responsible for ultimate acceptance of the project.

We have also made some suggestions for the process of standards development and distribution, to facilitate their application. We look forward to discussions around our findings and how we can work across industry to improve our approaches in future.

Acknowledgments The authors would like to thank the Directors of Ebeni Limited for allowing them to invest the time required to prepare this paper and present it to the Safety-Critical Systems Symposium 2023. We would also like to thank our colleagues who have supported us by providing constructive comments and suggestions.

Disclaimers The opinions expressed are those of the authors, based on their experience across a selection of industries where demonstration of safety is controlled by legislation and standards. We welcome feedback from our readers, especially if they wish to suggest improvements or extensions to the paper or insights into how the whole process can be improved.

References

[1] Health and Safety at Work Act, SI 1974/37, UK Government
[2] ALARP "at a glance", Health and Safety Executive (HSE) Website, https://www.hse.gov.uk/managing/theory/alarpglance.htm, Accessed 07 November 2022
[3] Management of Health and Safety at Work Regulations, SI 1999/3242, UK Government
[4] Common Safety Method for Risk Evaluation and Assessment, EU 402/2013 (as amended), Official Journal of the European Union, April 2013
[5] Railways and Other Guided Transport Systems (Safety) Regulations 2006, SI 2006/599, UK Government
[6] Construction (Design and Management) Regulations 2015, SI 2015/51, UK Government
[7] Common requirements for providers of air traffic management/air navigation services and other air traffic management network functions and their oversight, Commission Implementing Regulation (EU) 2017/373, https://www.legislation.gov.uk/eur/2017/373/contents, Accessed 9 November 2022
[8] Nuclear Installations Act, SI 1965/57, UK Government
[9] Railway Applications – The Specification and Demonstration of Reliability, Availability, Maintainability and Safety (RAMS), EN50126-1:2017 and EN50126-2:2017, CENELEC
[10] Railway Applications – Communication, signalling and processing systems – Software for railway control and protection systems, EN50128:2011[17], CENELEC
[11] Railway Applications – Communication, signalling and processing systems – Safety related electronic systems for signalling, EN50129:2018, CENELEC
[12] Railways Applications – Rolling stock applications – Software on Board Rolling Stock, EN50657:2017, CENELEC
[13] Functional safety of electrical/electronic/programmable electronic safety-related systems, IEC61508:2010, IEC[18]

---

[17] The latest version is actually addendum A2:2020.

[18] There are seven separate parts of this standard, and all are purchased separately.

[14] Information technology – security techniques – Information security management systems – Requirements, ISO/IEC 27001:2017, IEC[19],[20]
[15] Nuclear sites: Environmental Regulation, UK Government, https://www.gov.uk/guidance/nuclear-sites-environmental-regulation, Accessed 13 October 2022
[16] Cross-functional Software Standard for Railways, prEN 50716, CENELEC, 2022
[17] Rail Safety and Standards Board BowTie Hub, https://www.rssb.co.uk/en/safety-and-health/guidance-and-good-practice/bowties, Accessed 11 November 2022
[18] Guidelines and Methods for Conducting the Safety Assessment Process on Civil Airborne Systems and Equipment, ARP4761, SAE International, December 1996
[19] Design Assurance Guidance for Airborne Electronic Hardware, ED-80, EUROCAE, April 2000
[20] STPA Handbook, Nancy Leveson and John Thomas, http://psas.scripts.mit.edu/home/get_file.php?name=STPA_handbook.pdf, March 2018, Accessed 17 October 2022 (see also https://psas.scripts.mit.edu/home/)
[21] Service Assurance Guidance, SCSC-156B (version 3.0), SCSC SAWG, January 2022
[22] Safety of machinery. Safety-related parts of control systems - Part 1. General principles for design, ISO 13849-1:2020, ISO
[23] Safety of machinery: Functional safety of electrical, electronic and programmable electronic control systems, IEC 62061:2021, IEC
[24] Railway applications - Communication, signalling and processing systems - Application Guide for EN 50129 - Part 1: Cross-acceptance, CLC/TR 50506-1:2007, CENELEC
[25] Cross-acceptance of systems and equipment developed under different standards frameworks, IRSE News Issue 272, December 2020, Rod Muttram
[26] Safety decision making, S1521 A9, Transport for London, March 2019
[27] Railway applications - The Specification and Demonstration of Reliability, Availability, Maintainability and Safety (RAMS) - Part 5 - Functional Safety – Software, prEN 50126-5:2012, CENELEC
[28] Safety Assessment Principles for Nuclear Facilities, 2019/367414, Office for Nuclear Regulation, 2014 Edition Revision 1 (January 2020)
[29] "Design by committee" Wikipedia Entry, https://en.wikipedia.org/wiki/Design_by_committee, Accessed 20 October 2022
[30] Information Risk Management - HMG IA Standard Numbers 1 & 2, IS1&2 revision 4.0, CESG (now NCSC), April 2012
[31] Security and Privacy Controls for Information Systems and Organisations, NIST SP 800-53 Revision 5, NIST, October 2020
[32] Industrial communication networks. Network and system security, IEC 62443[21], IEC
[33] UK Acceptable Means of Compliance and Guidance Material for Regulation (EU) No 2017/373 as retained (and amended in UK domestic law) under the European Union (Withdrawal) Act 2018, 2017/373 ATM ANS AMC GM, CAA, December 2021, https://www.caa.co.uk/uk-regulations/aviation-safety/basic-regulation-the-implementing-rules-and-uk-caa-amc-gm-cs/atm-ans-provision-of-services/, Accessed 14 November 2022

---

[19] This is one of a series comprising 58 parts; again, all are purchased separately.

[20] This standard has been through several iterations (2005, 2013, 2017): where necessary the specific version being cited is clarified in the text.

[21] This is another standard published in many parts over a number of years: the introductory document (IEC TS 62443-1-1) is from 2009, with the latest part (IEC 62443-3-2, security risk assessment) dates from 2020.

Appendix A: Description of Legislation and Standards

The main legislation and standards mentioned in this paper are summarised below, for the benefit of readers who may not be familiar with these standards.

A.1 Common Safety Method for Risk Evaluation and Assessment (CSM-RA) [4]

This regulation defines the approach which must be applied to evaluate the safety of any "technical, operational or organisational" change to any mainline railway within an EU member state, and is therefore widely applied to railway projects of all types across the EU.

CSM-RA prescribes high level principles for safety management, but does not align this to any system development lifecycle. It focuses on the importance of hazard identification and management but does not have an explicit requirement for any form of safety case or safety justification. It does not prescribe any particular means of risk assessment and allows hazards to be managed by appeal to existing codes of practice, with explicit risk estimation allowed as a "fall-back" where existing codes of practice are not sufficient to manage the hazard. (Where codes of practice are applied, hazards do not need to be subject to risk assessment, although projects often (incorrectly) still apply a risk matrix to sentence the hazard.) There is therefore great pressure to justify that hazards can be managed using codes of practice, to avoid the effort and expense of explicit risk estimation.

The regulation does not set any specific targets (although it does prescribe a lower level on the numerical failure target which can be required from a technical system). It does not address the detailed analysis of technical systems at all.

Note: this is actually only one of several "CSM"s introduced under EU railway legislation, but it is the one most people mean when they say "CSM".

A.2 Railways and Other Guided Transport Systems (Safety) Regulations 2006 (ROGS) [5]

These are the underlying UK regulations for railway safety, which apply to all railway duty holders; they are relevant here because they define how the applicability of CSM-RA is determined in the UK and prescribe the arrangements for independent safety assessment in cases where CSM-RA does not apply (principally metros and other non-mainline railways).

A.3 Construction (Design and Management) Regulations 2015 (CDM) [6]

These are the UK regulations which manage health, safety and welfare relating to (building) construction projects (including new build, demolition, refurbishment, repair and maintenance); they include a requirement to consider risks relating to the use of the constructed edifice; application of these regulations tends to focus on immediate risks to people from the hazardous scenarios in the building(s) – so hazards arising from the functional use of the building(s) may not be considered (e.g. power exported by electricity substations or railway control actions implemented from a railway control centre). (CDM does include specific guidance for risks associated with specific types of construction work.)

Application of these regulations usually focuses more on "health and safety" risks, rather than the safety of use of a functional system. However there is a definite overlap – for example in assessing the safety of a station building.

Those applying CDM often undertake "Design Risk Assessments" although the term is not defined in the regulations, and no process for risk assessment is prescribed. As a result, there is generally no in-depth analysis of the risks, just a description of risk and mitigation.

A.4 Nuclear Installations Act 1965 [8]

Non-military nuclear sites in the UK must comply with this act, which has three key purposes: it defines a licensing scheme for nuclear sites; it defines a permit system controlling enrichment and extraction of nuclear material; it defines legal liabilities of licensees towards third parties. The licensee must demonstrate that they maintain control and oversight of safety on the site at all times, and there is a series of licence conditions which provides more detail on what is required.

A.5 Implementing Regulation 2017/373 (Air Traffic) [7]

Providers of Air Traffic Management (ATM) and Air Navigation Services (ANS) and related supporting services in the UK must comply with this legislation and its amending Statutory Instruments.

The Regulation lays down common requirements for providers of ATM and ANS and other air traffic management network functions and their oversight. The ATM/ANS IR is based on ATM-related ICAO Standards, Recommended Practices (SARPs) and Procedures for Air Navigation Services (PANS).

Under this legislation, providers of Air Traffic Services (ATS) must carry out safety assessment of changes, and providers of other supporting services including Communication, Navigation and Surveillance (CNS), Aeronautical Information Services (AIS), Meteorological Services (MET) and Airspace Design Services must carry out safety support assessment of changes to the service, providing assurance, with sufficient confidence that the service will behave and will continue to behave only as specified in the specified context.

The CAA's guidance material [33] for both safety assessment and safety support assessment references a number of industry standards and guidelines relating to software safety, primarily those produced by EUROCAE and RTCA, but also IEC 61508 part 3.

A.6 CENELEC EN 50126 [9]

This is the standard for assuring reliability, availability, maintainability and safety in the development of railway technical systems. It is widely used for railway signalling system development and is gaining increasing recognition for other technical systems on the railway. The standard is strongly based on a "V-model" lifecycle, and defines the inputs, outputs and activities to the undertaken at each lifecycle phase. It also defines roles and organisation structure constraints, and introduces the concepts of hazard analysis, risk assessment and SILs. It aligns to CSM-RA in specifying three risk acceptance principles and introduces the concept of a safety case and provides guidance on its structure and content. There is also guidance on safety analysis techniques and derivation of safety targets.

A.7 CENELEC EN 50128 [10] / EN 50657 [12]

This is the standard for development of safety critical software for railway control and protection systems: EN 50128 and EN 50657 are for infrastructure and rolling stock respectively and are very similar although there are differences in the requirements for independent assessment at low integrity levels.

These standards are dependent on EN 50126 to define the safety requirements for the software. They define lifecycle and process for software development with inputs, outputs and activities to be undertaken at each stage of the lifecycle. They provide requirements for techniques to be applied at each lifecycle stage, dependent on the SIL assigned to the software functions and include a long catalogue describing these techniques in more detail.

In addition, they provide some coverage for extending the principles of the standard to the development of configuration data.

A.8 CENELEC EN 50129 [11]

Often thought of as the standard for railway safety cases, this goes into more detail (than EN 50126) on the structure and content of safety cases as well as the derivation and apportionment of failure targets; it also has a lot of guidance on the hardware of safety critical systems, including hardware failure modes, hardware design architectures and techniques.

A.9 IEC 61508 [13]

This is a basic functional safety standard applicable to all industries for the development of electrical / electronic / programmable electronic safety-related systems. Many industries have produced derivative standards which adapt the principles from IEC 61508 to be more appropriate to their own practices. The standard defines a safety lifecycle for the development of such systems and guidance on how to evaluate the risk associated with the operation of the system and how to reduce this risk to an acceptable level. It includes specific techniques for use at each phase of the lifecycle, dependent on the level of risk.

A.10 ISO 27001 [14]

This is the most commonly applied standard for the security of information technology systems. It lists a set of requirements for establishing and maintaining an information security management system at an organisational level to protect the confidentiality, integrity and availability of its data. This is achieved by applying a risk based approach which is managed throughout the life of the system. The set of baseline controls covers, processes, organisation (including leadership), technology and information. An organisation which: (a) implements these requirements in full and (b) is audited to ensure their implementation is effective is viewed as having good information security.

A.11 IS 1&2 [30]

This standard was originally created in 1998/1999 as two separate documents as part of the UK Government's Security Policy Framework (SPF). It was updated and managed by CESG (the former name for NCSC). It presented a set of twenty

Risk Management Requirements for organisations to implement and mandated them across central governmental organisations as the basis of their information risk management policy. This standard also included a supplement which details the risk management process which should be applied along with a baseline control set which was tailored from ISO 27001:2005. Control implementation guidance is tailored from ISO 27002:2005 to provide three different levels of implementation: Deter, Detect and Resist and Defend. Since the 2014 update to the SPF, IS1&2 is no longer mandated, and the decision was taken by CESG to no longer maintain the document. It has been retained as a legacy document with the final update being conducted in 2012.

A.12 Safety Assessment Principles (SAPs) for Nuclear Facilities [28]

The ONR's SAPs are used by inspectors to guide regulatory decision making in the nuclear permissioning process. Underpinning such decisions is the legal requirement on nuclear site licensees to reduce risks so far as is reasonably practicable. The SAPs apply to assessments of safety at existing or proposed UK nuclear facilities, applying (only) nuclear safety, radiation protection and radioactive waste management. Conventional hazards associated with a nuclear facility are excluded. The primary purpose of the SAPs is to provide inspectors with a framework for making consistent regulatory judgements on the safety of activities. The principles are supported by Technical Assessment Guides (TAGs), and other guidance, to further assist decision making within the nuclear safety regulatory process. Although it is not their prime purpose, the SAPs may also provide guidance to designers and duty-holders on the appropriate content of safety cases.

The principles are grouped into the following categories:

- Fundamental Principles
- Leadership and Management for Safety
- Regulatory Assessment
- Siting
- Engineering Principles (with multiple further subdivisions)
- Radiation Protection
- Fault Analysis
- Numerical Targets and Legal Limits
- Accident Management and Emergency Preparedness
- Radioactive Waste Management
- Decommissioning
- Land Quality Management

The SAPs are used by the Licensee to support their own safety assessments and the development of their own arrangements and principles.

Case Study Analysis of STPA on an Industrial Cooperative Robot and an Autonomous Mobile Robot

Laure Buysse[1], Simon Whiteley[3], Jens Vankeirsbilck[2], Dries Vanoost[1], Jeroen Boydens[2] and Davy Pissoort[1]

[1] KU Leuven, Faculty of Engineering Technology, Department of Electrical Engineering

[2] KU Leuven, Faculty of Engineering Technology, Department of Computer Science

[3] Whiteley Aerospace Safety Engineering & Management Limited

Abstract *Autonomous systems are becoming more and more prevalent within industry. However, it is no easy feat to ensure their safety. Current safety approaches struggle to deal with the unprecedented levels of complexity introduced due to these new autonomous systems. More recently, System-Theoretic Process Analysis (STPA) was introduced to help solve some of these problems. However, real-world examples or practical reviews and guidance are still hard to find. To bridge this knowledge gap, this paper takes a critical look at STPA using the results of two case studies: an autonomous and a collaborative system. We present a loss and hazard list for autonomous mobile systems, alongside a more systematic method to structure certain steps within the analysis. Additionally, we reflect on the challenges that had to be overcome and highlight the differences between applying STPA on new systems, as opposed to applying STPA on existing systems. We highlight the importance of using the correct language / vocabulary and discuss how to build confidence in the results achieved by performing STPA*

1 Introduction

Automated and autonomous systems are becoming more and more prevalent. From search robots in factories to the development of self-driving cars, new technology and breakthroughs happen almost every day (Zhang et al., 2017). However, these new technologies and scenarios (black-box deep learning models, complex, unknown environments and self-learning systems) presents an unprecedented level of complexity. It is increasingly difficult to ensure the safety of these systems (Burton et al., 2020). Current safety approaches often struggle to analyse and manage these new technologies. Moreover, while there is a huge

© KU Leuven 2023. © Whiteley Aerospace Safety Engineering & Management Ltd 2023.
Published by the Safety-Critical Systems Club. All Rights Reserved.

benefit to shift system analysis to earlier stages of the design and development cycle (Boehm, 1984), traditional methods tend to require detailed diagrams and component lists which are not yet available at that point in time (PQRI, 2015) (Siemens, 2016).

More recently, STPA was introduced to help solve these problems. STPA treats safety as a dynamic control problem, no longer focussing on chain-of-failure-events, (Leveson and Thomas, 2018). Due to the modelling similarities between an STPA control structure and models classically used for autonomous systems, such as OODA[1] or SUDA[2], STPA seems an excellent starting point for early and complete analysis of these systems (Buysse et al., 2022). However, real-world examples and practical guidance or reviews on the application of STPA are still hard to find.

To bridge this knowledge gap and as a foundation for discussion, this paper takes a critical look at STPA using the analyses performed on two case studies. We provide detailed outcome from the case study analyses along with some practical guidance. Additionally, we review the results and the process, and discuss topics such as the difference between analysing existing or new systems, expected results, the importance of using the correct terms and tool support.

The remainder of this paper is structured as follows. Section 2 briefly introduces System-Theoretic Accident Model and Processes (STAMP) as well as System-Theoretic Process Analysis (STPA). The case studies are described in Section 3. The results of both analyses are described in Section 4. In Section 5, we provide a critical reflection on the STPA results and the process of applying it. Finally, Section 6 presents the conclusion.

2 STAMP and STPA

As introduced by Leveson et. al., Systems-Theoretic Accident Model and Process (STAMP) is the name of an accident causality model rooted in systems theory (Leveson and Thomas, 2018). Unlike more classic hazard analysis techniques such as HAZOP or FMEA, STAMP considers safety as a system control problem. Rather than studying the chain of directly related failure events or component failures, STAMP looks at system interactions defined within a control structure (Leveson and Thomas, 2018).

---

[1] The 'Observe-Orient-Decide-Act'-model (OODA) is a four-step approach to decision-making. It focuses on filtering the information, putting it in context and making the most appropriate decision.

[2] SUDA (Sense-Understand-Decide-Act) is an agent model for robotics and autonomous systems (RAS) consisting of four main elements. Among other things, the model is used to aid the definition, implementation and validation of RAS system requirements.

System-Theoretic Process Analysis (STPA) is a hazard analysis technique which builds upon this model, identifying possibly dangerous situations using control actions within a control structure. Additionally, constraints and requirements are defined at different steps throughout the process (Leveson and Thomas, 2018).

In theory, the STPA process is simple, consisting of four major steps. For completeness, we define two additional steps, namely "verification and validation of the results and the STPA process" and "update of the analysis". Table 1 gives a quick overview of all steps and their main goals.

Table 1. Steps within the STPA process and their associated goals (adapted from Leveson et. al. (Leveson & Thomas, 2018)).

Step	Name	Goal
1	Define	- define (stakeholder) [losses][3]
		- define [hazards]
		- define system boundaries
2	Model	- model the abstract system control structure
		- refine substructures as desired / needed
3	Identify unsafe actions	- explicitly define the context
		- analyse all control actions within the system control structure(s)
4	Identify causal scenarios	- identify the causes for all unsafe control actions
		- pose solutions for all identified issues
(5)	Verify and validate the results / STPA process	- verify and validate both the results and the STPA process itself using domain experts
(6)	Update	- update the analysis based upon design changes, review comments ...

3 Case Study Description

The case studies used within this paper are based on two industrial grade demonstrators. The first is a classic autonomous system is analysed, i.e. an autonomous mobile robot (AMR). As a second case study, a more complex system is examined: a large industrial robot which is to be converted to a (partial) autonomous, partially collaborative system. Our first case study details the analysis of an already existing system. The second case study starts from an earlier point in the design-cycle, where most details are still unknown.

---

[3] For clarity, square brackets are used around specifics terms used within STPA as defined by Leveson et. al. in the STPA Handbook (Leveson and Thomas, 2018).

3.1 First Case Study for Analysis: Autonomous Mobile Robot

AMRs are frequently used within (autonomous) warehouses to efficiently and safely move workloads and more. They offer a scalable and customisable solution to automate many processes. Due to its prevalence in the industry, this case study is an ideal baseline to evaluate STPA. The case study in question features an AMR, which is used within a factory setting to transport packages. Part of a fleet of three, the robots receive tasks from a central server and plan out their own routes accordingly. Each task follows the same pattern:

(1) Calculate a path and drive to location A,
(2) Dock at a specific module,
(3) Pick up the package,
(4) Drive a predetermined path to location B,
(5) Dock at the specific module,
(6) Drop off the package

After a task is successfully completed, the above process is either repeated or the AMR automatically drives to a designated charging station and switches to standby mode. This decision is made based upon current battery life, the total workload and timing constraints. Needless to say, every task should take place as safely and efficiently as possible, keeping in mind the humans moving around within the AMRs environment and the factory infrastructure itself.

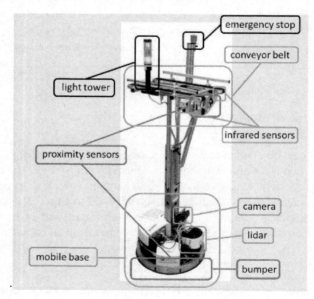

Fig. 1. Case study 1 – autonomous mobile robot

The basic design of the already existing system, detailed in Figure 1, consists of a moving platform with three omni-directional wheels, a bumper, two batteries and a control unit. Transport is made possible using a tower with a conveyor belt. For navigation purposes, the robot is equipped with a camera, a LIDAR[4] and an array of proximity sensors (on the bumper and the transport tower). Additionally, a sensor is available to detect the presence or absence of a payload. A light tower is used to communicate the system state to the outside world. Communication to the docking stations (for pick up and drop off) takes places using infrared (IR) sensors.

3.2 Second Case Study for Analysis : Large Industrial Cooperative Robot (ICR)

Cooperative robots (cobots) are a quickly emerging technology. Most often deployed within manufacturing and assembly industries, cobots are designed to operate alongside and interact with humans in a shared workspace (Hentout et. al., 2019). Typically these systems consist of a small fixed robot arm, which is able to rotate and move within a fixed space. A variety of end-effectors can be used for different applications such as welding, drilling, transport etc.

The second case study within this paper aims to take the concept of the cobot one step further. Used to aid the inspection of heavy workpieces, a heavy, large industrial robot is to be deployed to function as a cobot (Figure 2). This human-robot collaboration setup or industrial cooperative robot (ICR) is to assist a human operator, where the robot takes on the tasks which require strength and the human takes on those which require dexterity and mental capacity. The operator is in charge of inspecting workload by manually manipulating the robot within the collaborative zone. The robot needs to autonomously pick up the pieces, present them to the operator and put them back down in the designated spots when required.

---

[4] LIDAR (LIght Detection And Ranging of Laser Imaging Detection And Ranging) is a method for determining ranges (variable distance) by targeting an object or a surface with a laser and measuring the time for the reflected light to return to the receiver.

Fig. 2. Case study two: the industrial cooperative robot (ICR)

To accomplish the task above, the concept details three LIDAR sensors, placed around the robot to scan the environment. The back of the robot, which does not have such a sensor, should be fenced off and thus cannot be reached. The boundaries of the collaborative and autonomous workspaces are currently projected onto the floor to assist the user. Additionally, the design details a light tower at the bottom of the robot. The robot itself is currently fitted with some sensors, including force-torque sensors, intended for navigation and localisation.

4 Application of STPA

When starting the analysis, it is crucial to start by explicitly defining all basic system properties and assumptions. For a complex system, this is even more important as faults are easily introduced when the many requirements and assumptions are not well documented or not known for all stakeholders within the analysis. The following elements should always be defined before starting:

(1) Well-defined and clear system boundaries
(2) A list detailing all assumptions
(3) A list of (possible) stakeholders such as users, customers, regulatory organisations ...
(4) A list of all requirements set by these stakeholders

Thereafter, the analysis can begin by defining the system [losses]. An example [loss]-list for both the AMR and the ICR can be found in Table 2. Though both systems have quite different designs, their loss lists are identical. This similarity can be found with most systems with similar task (e.g., transport) within a similar domain. While additional losses, based on different stakeholders, regulations or specific system environments, can differentiate the list, certain [losses] will be important for most systems. What's more, a baseline list should not be excessive. Generally, seven to ten [losses] will suffice. When a larger amount of [losses] does present itself, it might be helpful to (1) bundle [losses] together using a more general term or (2) redefine system boundaries more concretely.

Table 2. Identified [losses] for the two use cases

	AMR	ICR
[L-1.1]	Human death	
[L-1.2]	Human injury	
[L-2]	Loss of or damage to the robot	
[L-3]	Other asset or equipment damage	
[L-4]	Mission disruption	
([L-5])	(Environmental loss)	

Next, the system-level [hazards] should be defined and coupled to their respective [losses]. Table 3 documents the hazards for both case studies. Once more, there are quite some similarities between both lists. Once again, this can be attributed to the similarity in the system assignments (transport of goods). STPA is a technique focussed on system interactions, functionality and emergent properties rather than physical design. As a consequence, similar [hazards] should be expected in systems with similar functionalities. Moreover, the additional [hazards], as defined for the ICR, are exclusively related to its cooperative functionality, a function not found in the AMR case study, further confirming our hypothesis.

In spite of these similarities, defining [hazards] within STPA is not quite as easy as one would hope. While current standards and user manuals often provide lists, these cannot be used directly within STPA due to a definition mismatch. More concretely, current norms such as ISO 12100:2010 define a hazard as "*a potential source of harm*" (ISO, 2010). STPA on the other hand defines a [hazard] as "*a system state or set of conditions that, together with a particular set of worst-case environment conditions, will lead to an accident (loss)*". Currently, there are no STPA [hazard] lists available. As a result, establishing all possibilities can be quite difficult and tedious the first time around.

Table 3. Identified [hazards] for the AMR and ICR linked to the losses presented in Table 1

	AMR	ICR	Loss
[H-1]	The system loses its load.		[L-1.2] [L-3] [L-4]
[H-2]	The system violates the minimum separation rules between the system and other objects or people		[L-1.2] [L-2] [L-3]
[H-3]	The system collides with objects, people or other infrastructure		[L-1.1] [L-1.2] [L-2] [L-3]
[H-4]	The system entraps objects, people or itself.		[L-1.1] [L-1.2] [L-2] [L-3] [L-4]
[H-5]	Interactions and behaviour which cause mission delays		[L-4]
[H-6]	The system is under excessive stress		[L-2] [L-3] [L-4]
[H-7]		The system causes human exposure to extreme temperatures	[L-1.2]
[H-8]		The system cause human exposure to electrical power	[L-1.1] [L-1.2]

At last, the first stage of the analysis is finalised by specifying [high-level safety requirements] ([HLSR]) for each identified [hazard]. In general, there are four main goals identified at this level (Whiteley, 2022a):

(1) Complete [hazard] elimination
(2) [Hazard] prevention
(3) Protection of the environment / people / infrastructure
(4) [Hazard] response / reaction after-the-fact

Multiple [HLSR] can be defined for each [hazard], typically fitting different categories. A single requirement can also trace back to multiple hazards (Leveson and Thomas, 2018). Table 4 defines the [HLSR] for *[H-2]* and *[H-3]* as defined in Table 3.

Table 4. [High-level safety requirements] for [hazard] 2 ('The system violates the minimum separation rules between the system and other objects or people') and [hazard] 3 ('The system collides with objects, people or other infrastructure')

Hazard	Requirements	Category
[H-2]	The system must adhere to the minimum separation rules from other objects and people.	Eliminate
	If the minimum separation rules are violated, this must be detected and additional measures need to be taken to prevent property damage and/or collisions.	React

Hazard	Requirements	Category
[H-3]	The system shall stop immediately after a collision	React
	The system shall be able to withstand minor collisions	Protect
	The system must adhere to minimum separation rules form other objects and people (see also [H-2])	Eliminate

Modelling a system control structure is the second stage of the STPA process. Based upon a simple control loop, a hierarchical control structure is designed to model all system behaviour (Leveson and Thomas, 2018). It is important to note that an STPA control structure is neither a physical (hardware) model nor an executable model of the system in question. It is purely a functional model and should be designed as such. To help manage complexity, model abstraction should be used. Moreover, it is normal to have abstract models at the start of the design process where many details are still unknown. Later in the design, details can be added to the model. For autonomous systems, it is beneficial to design a single abstract system model accompanied with multiple detailed models each highlighting a specific sub-system. As such the abstract model provides a system-wide overview, while the detailed models can be used for more detailed analyses of specific functionalities.

Paying sufficient attention to this step is vital for a good analysis. Not only does the model form the basis for the rest of the analysis, often fatal flaws, such as single points of failure or missing feedback, are already identified in this early stage of the analysis. For example, Figure 3 details a higher-level control structure for the ICR. Here it can be seen that feedback is missing from the warning system (lights) to the central system controller. In other words, the control loop is not closed. Thus, currently the ICR is unable to check if the warning system is functioning correctly at all, which is a of course a huge safety concern.

Once all properties are defined and the system is modelled, the analysis can proceed to identify [unsafe control actions] ([UCAs]). STPA defines four ways any [control action] ([CA]) in the control structure can be unsafe (Leveson and Thomas, 2018):

(1) *Provided:* The action is provided when it should not have been.
(2) *Not provided:* The action is not provided when it should have been.
(3) *Timing:* The action is provided when needed but is coming too late, too early or in the wrong order / sequence.
(4) *Duration:* The action is provided too long or too short (continuous signals only).

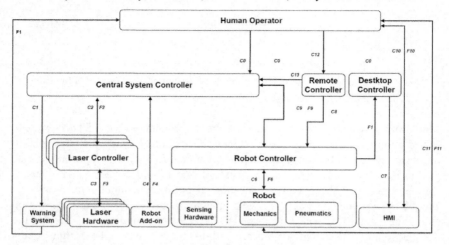

Fig. 3. High-level control structure for the ICR. Arrows leading down represent a set of control actions (labelled Cx). Arrow leading up represent the feedback given (labelled Fx). Where both control actions and feedback are present, double arrows are used. The blocks within the diagram represent the high-level controllers. Multiple blocks on top of each other, indicate the presence of multiple identical controllers. Blocks within the controller detail more information which can then be used in lower-level control structure.

While these categories provide a solid foundation to start generating [UCAs], for autonomous systems this step is quite daunting due to the sheer number of possible scenarios and issues. Moreover, while the STPA handbook acknowledges the context is critical for each [UCA], it does not provide guidance around scenario management or generation. For complex systems, we recommend the use of multiple context tables. A generic context table can be used to highlight generic system-states (e.g. emergency mode), which need to be considered for every [CA]. Task specific or controller specific tables can be used for more detailed contexts only applicable for a specific task or controller.

Systematically structuring the possible contexts will not only help to structure the analysis, but will also give more confidence in the completeness of the analysis itself. Moreover, as all [UCA] follow the same phrasing, possible scenarios could be generated partially automatic using the [CAs] within the control structure and context lists (Table 5). Of course, the safety team will still have to judge each action as "wanted behaviour" or "unsafe" and attach the associated [hazard(s)] if the action falls into the latter category. We do warn against automation within this process. While it can be a helpful tool, it can also lead to complacency and misidentified or undetected scenarios. Moreover, for beginners manually going through the process helps build the critical mindset vital to the process.

While studying possible [UCAs], it must be said that more often than not, most unsafe scenarios might be known or obvious to the practitioner. However, it is good practice to explicitly document these scenarios. Moreover, explicitly documenting all identified scenarios can help reduced the number of dangerous scenarios which are overlooked, i.e. contexts which are not considered or situations which are more dangerous than initially thought. Moreover, by considering and analysing more obvious scenarios, it will become easier to analyse and implement system changes in the future.

Table 5. The standard format for a [UCA] is presented within this table. Beneath an example [UCA], which complies with the standard, is given from the ICR analysis. A shorthand version of the example is presented on the last line. Within this format, 'C8.1' corresponds to the control action under analysis, '[0]' represent the fault category, 'CT.C1' links the specific context which is considered. Finally, a verdict ([!] = hazardous or [OK] = normal) is made and the hazard(s) to which the [UCA] might lead are linked ([H-5]).

Standard form	<[Control action]> <Category> <Context> * <Label> * <Hazard>
Example	<Move robot arm> <not provided> <when a workpiece is ready to be transported> * <Unwanted> * <[H-5]
Shorthand	<C8.1><[0]><CT.C1> * <[!]> * [H-5]

Based upon the [UCA], another set of requirements should be defined. In essence, these are the negative of the [UCA] itself. Table 6 provides an extract from the ICR [UCA]-table (based on the tables from (Whiteley, 2022b)). The structure of the table is designed with human factors and human performance in mind. The task at hand (identifying [UCAs]) is horizontally split into three distinct elements: problem definition, safety judgement and solution definition. Vertically, all the different contexts and scenarios are presented. While doing the analysis, it is up to the practitioners to decide whether to go down the list and define all the problems first or take it one scenario at a time and both define and judge the scenario, providing safety requirements along the way.

One example [UCA] is provided within each category for one given [CA]. Diverse contexts are used to illustrate multiple hazard links and scenarios. In practice, every table should contain the analysis regarding one [CA] with respect to only a single context or a set of appropriately bundled contexts to maintain a structure and clear analysis.

Lastly, [causal scenarios] ([CS]) should be derived from the [unsafe control actions]. Four non-exhaustive categories are defined within the STPA theory. Each category is linked to a specific part of the standard control loop which forms the basis of the entire analysis.

(1) Unsafe controller behaviour
(2) Inadequate feedback / data / information
(3) Control path issues
(4) Factors related to the process

Generally, we recommend defining four tables (corresponding to the four categories) for each control loop in question. This will help with clarity and structure as well as traceability throughout the analysis. Each scenario should also be followed up with at least one requirement. These requirements will usually provide detailed information regarding implementation or testing.

Table 6. An example UCA table (as adapted from Whiteley S. (Whiteley, 2022b)) is presented for a single [control action] (opening the gripper of the ICR). The [control action] is analysed based around different contexts for each fault category to illustrate different [hazard] links and non-hazardous versus hazardous outcomes.

Type	Problem Statement — Control Action [CA] – Open the gripper of the ICR	Judgement — Unsafe control action? Hazard link [H]	Solution Statement — Controller requirement
Not provided [0]	[x] does not provide [CA] when [Context] [robot controller] does not provide [open] when [a new package is ready for pickup]	[H-5]	[x] must provide [CA] when [context] [robot controller] must provide [open] when [a new package is ready for pickup]
Provided [1]	[x] provides [CA] when [Context] [robot controller] provides [open] when [the robot is executing transport]	[H-1]	[x] must not provide [CA] when [context] [robot controller] must not provide [open] when [the robot is executing transport]
Timing early [TE]	[x] provides [CA] [early] when [Context] [robot controller] provides [open] [early] when [the drop-off area isn't reached yet]	[H-1]	[x] must not provide [CA] [early] when [context] [robot controller] must not provide [open] [early] when [the drop-off area isn't reached yet]
Timing late [TL]	[x] provides [CA] [late] when [Context] [robot controller] provides [open] [late] when [a new package is ready for pickup]	[H-5]	[x] must not provide [CA] [late] when [context] [robot controller] must not provide [open] [late] when [a new package is ready for pickup]
Timing sequence [S]	[x] provides [CA] [in order y] when [Context] [robot controller] provides [open] [after close] when [picking a package up]	[H-6]	[x] must not provide [CA] [in order y] when [context] [robot controller] must not provide [open] [after close] when [picking a package up]
Duration short [DS]	[x] provides [CA] [too short] when [Context] [robot controller] provides [open] [too short] when [picking a package up]	[H-3] [H-6]	[x] must not provide [CA] [too short] when [context] [robot controller] must not provide [open] [too short] when [picking a package up]
Duration long [DL]	[x] provides [CA] [too long] when [Context] [robot controller] provides [open] [too long] when [picking a package up]	Not hazardous	[x] must not provide [CA] [too long] when [context]

5 Critical Reflection – The Pathway to a Good Analysis

This section contains reflections on the challenges experienced with learning, practicing and getting results with STPA, along with details on how these challenges were embraced / overcome.

A number of these challenges are not strictly due to the STPA process itself, but had more to do with developing understanding and skills in how to apply STPA in the specific context of interest, i.e. existing systems.

5.1 The Importance of an Open Learning Culture

Broadly speaking, it is important to recognise that when attempting to apply any new approach, the personal journey of learning and gaining experience involves a challenging period of trying to understand the language, vocabulary, concepts and the overall flow of that new approach, along with developing related thinking and practical skills.

Of course, individual understanding, skills and results won't be flawless from day one, and mistakes / failures are a necessary part of that journey and must be celebrated and embraced.

Developing and embracing an open learning culture, as part of a broader "Just Culture" (Dekker, 2012), is fundamental to a successful application of STAMP / STPA and, most importantly, benefiting from the results gained.

This is especially important where sensitive issues, including potential safety issues, are identified which require careful, open and respectful consideration to resolve.

5.2 Analysing Existing Systems

The STAMP / STPA learning journey, and attempts to get results, is made all the more challenging when attempting to learn and apply the new approach to an existing system, or a change to an existing system. It is also made more difficult where conversations with the system's originator are not possible.

This is because individuals are not just 1) learning and developing a new skill and exploring STPA, they are, broadly speaking, also 2) attempting to learn about and make explicit a set of problems that are considered, by the system originator, to have already been solved by the existing system as implemented.

This second aspect can be overwhelming and lead to an incredible amount of uncertainty on the part of the individuals, which can undermine the speed at which they learn and gain experience and confidence in STPA and in their abilities to apply it.

Speaking more generally, this is further complicated where detailed information about the (existing) system is limited or not available, for whatever reason.

There also exists the potential for individuals applying STPA to overlook important issues that are not apparent given their experience with the existing system but also, and conversely, they may identify issues that may not appear to have been resolved by the existing design, i.e., issues that were previously hidden, unrecognised, or even suppressed.

These "new issues" may simply be due to the observer's state of knowledge and experience, or they could be genuine issues with the existing system, but without any kind of conversation with the system originators, it is difficult to conclude if the issues identified are genuine or just issues with perceptions / perspective. This can lead to a level of uncertainty which can affect individual confidence in their skills, along with slowing overall analysis progress.

Generally speaking, and regardless of whether the issues are genuine or not, conversations with system originators / operators can be rather difficult if they don't have an open learning culture and willingness to embrace STPA results. This could be compounded where an individual's confidence is being undermined.

5.3 Analysing New Systems

This is in stark contrast to the level of effort and complexity involved in attempting to learn and apply STPA to a new system (or concept), where the safety problems and their solutions have yet to be fully defined, and implementation of those solutions has yet to take place.

Ultimately, learning to apply STPA to a new system is less of an overhead / burden on individuals, though there is still tremendous value to be gained from applying it to existing systems; it's just that the organisation / individuals involved have to be respectful of the extra level of challenge involved, especially where limited detailed information about system design, requirements, safety assessments, implementation etc. are available.

Fundamentally, regardless how great the analysts or their STPA results are, if the organisation in question is not set up to adequately respond to the STPA findings, then it does not matter how good the analysis results actually are. It will not be possible to adequately influence the system design or operation.

5.4 STPA early application: managing expectations

As part of the initial [control action] analysis activity (immediately after control structure model creation and development), it was highlighted that a significant proportion of early results were considered obvious, i.e., that certain [unsafe control actions] could be identified through logical inference and experience, especially when related to known failure modes.

This initial experience raised reasonable questions about justifying the value of STPA beyond existing approaches because it was highlighting things that were already known / suspected.

This experience had the potential to undermine people's initial confidence and willingness to invest further resources in learning and applying STPA, especially where there was a not unreasonable view that STPA should immediately demonstrate its value by finding new things that are not already known or covered by other failure-based approaches.

Consequently, it is important to manage stakeholder expectations and ask them to withhold judgment until after the first few hours of application have passed (typically about two hours beyond control structure modelling) at which point the subtle and significant issues start to emerge, often in the form of "dangerous successes", i.e. the system did not fail, it behaved as designed and that has the potential to result in a [hazard] / [loss].

The time taken to gain powerful results and evidence that justifies the investment in STPA varies, but is typically within the first two hours of control analysis activity, after the control structure model has been developed.

5.4.1 Confidence Building

It is important to recognise that working systematically through the STPA process essentially provides the opportunity to confirm (and capture) those "obvious" findings, i.e., what is already considered as known, and helps to clarify and capture important context.

This confirmatory aspect of STPA helps to build a level of confidence in the existing system design and previous safety assessments, but also tease out the not so obvious potential dangers hidden within that same design.

When the analysts start to get comfortable with the STPA process flow and the flow of their thoughts / conversations, the identification of [unsafe control actions] begins to flow thick and fast, at which point it becomes glaringly apparent that those really subtle but significant findings are, broadly, more common than initially expected.

This realisation can be rather disconcerting at first, but, conversely, rather comforting because the issues have been found and have been turned from previously "unknown unknowns" to "knows unknowns" and, as more information is made available, eventually transformed into "known knowns". This experience helps to build analysts' confidence in their own skills, but also in the STPA process itself.

5.5 Creating the Mindset: Disciplined Speech and Thought

To be able to perform a good STPA analysis and gain valuable results, individuals require discipline in their use of STAMP / STPA language, vocabulary and concepts. This is especially important when talking with different domain-experts and/or practitioners about STPA / STAMP concepts and other safety assessment approaches or safety management concepts, such as when discussing the concept of a "hazard" / [hazard].

The need for disciplined speech and thought often comes across, at least initially, as being rather pedantic and constraining. However, as time elapses and experience builds, it becomes apparent how critical that discipline is for individuals' clarity of thought, discussion, problem definition, solution definition and control structure modelling.

This is especially true when creating / developing control structure models and moving conceptually between the physical and functional representations of controllers and their interactions, as represented by the control structure model.

5.5.1 One's Own Tongue

There is often a motivation to translate STAMP / STPA language and terminology into one's own technical / organisation / domain-specific vernacular. This is completely understandable, but should be respectfully resisted for two key reasons.

First, any attempt to translate and then communicate using the translated form requires extra effort, overhead and complexity that requires significantly more resources, energy and time to accommodate, which can lead to expensive misunderstandings and mistakes that can have very real safety consequences. It also means that people who are to be involved in STAMP activities will require extra training and time to develop their skills, over and above necessary key skills. Ultimately, the need to translate the language / terminology presents a barrier to adoption and can cause the application of STPA to become really complex very quickly causing unnecessary confusion and frustration. This undermines efforts to gain benefits from STPA, which could be very costly, and potentially dangerous.

Second, learning and using the published STAMP / STPA language and terminology means that analysts can apply their skills and experience to any type of system, and not just those that exist in the industry that they currently support. It also means that experience can be shared across industries.

5.6 STPA as a Team Sport

STPA analysis is a team sport. It is important that key team members have some level of experience in systems engineering and system safety engineering processes and application, including requirements engineering. The STPA process would typically be facilitated by system safety engineers who would be supported by a project team of people who represent different disciplines, and who also represent the various controllers and interactions within the control structure model, e.g. operators, human-machine interface design engineers, human factors engineers / psychologists, etc. Given the context of learning and applying STPA to an existing system (e.g. the AMR) with relatively superficial level of technical detail, STPA step 4, the [casual scenario] analysis was rather challenging, because little detail was known about the physical implementation of each part of the control loops.

This was unfortunately unavoidable and made practicing [causal scenario] analysis rather limited. However, it was recognised that had the analysts been working in collaboration with the robot designer / manufacturer, and had access to technical detail and relevant electrical, mechanical and software engineers, this part of the analysis could have been done to a much deeper level of detail, with associated specific safety constraints being defined.

For example, when exploring potential [causal scenarios] influencing controller behaviour, environmental conditions including temperature and electromagnetic interference (EMI) were considered as potential contributors, and generic safety constraints regarding environmental conditions were considered for specification.

However, had there been an option to collaborate with the controller supplier / designer / manufacturer, questions regarding environmental qualification could have been asked with the expectation that appropriate design and implementation had been performed with relevant qualification test evidence.

This experience emphasised the critical need for the STPA analysis activity to be performed by a team that either: 1) is made up of stakeholders who have knowledge, experience and information that represent the controllers and interactions represented in the control structure model, or 2) has access to those stakeholders / necessary information.

Another important suggestion is to involve test engineers / members of the system test organisation because their prime objective is to try and break the system. They are typically able to think very creatively about how things might not behave as intended, or how they can be encouraged to misbehave.

5.7 Tool support: *From Pen and Paper to Word, Excel, Visio, and Beyond!*

When starting to learn and practice STPA analysis, it is absolutely essential that the analysts perform the activity by hand, i.e. using a simple approach involving pen, paper, post-it notes and a whiteboard (or their digital equivalents), with participants gathered around together physically (or virtually) as a mastermind group. This is so they can exploit the opportunity to develop the necessary thinking, speaking and practical skills necessary for a good STPA analysis.

However, as the complexity of the control structure model increases and as the number of controllers, their interactions (especially [control actions]), system contexts and analysis stakeholders increases, it logically follows that the complexity of the analysis, and its conduct, increases. Consequently, it is recommended that template documents or forms are used to help facilitate the systematic performance and documentation of the analysis, whilst taking care to use an appropriate version / configuration control of any information / documentation used as inputs and the resulting analysis outputs, including the control structure model(s).

When attempting to apply STPA to a complex system / complex control structure model, some level of software tool support, such as MS word, Visio, MS Excel, PowerPoint or Miro, is recommended so as to reduce the cognitive workload on the analysis participants when trying to recall and remember lots of detailed information, and ensure traceability / consistency of the analysis.

5.7.1 Beware Automation and Software Tool Support

However, when graduating to a level that needs software tool support for more complex applications, care must be taken not to automate the analysis to such an extent that it removes / degrades the creative thought process that is absolutely fundamental to a good quality analysis.

Graduating to using software tool support too soon, or too extensively, can lead to the analysts thinking and application skills being underdeveloped, whilst they themselves may not realise their own limitations. As humans are naturally adept at optimising for minimal energy use, this can lead to a level of complacency in the conduct, or documentation, of the analysis. This means that things can be missed, and thinking and practical skills do not develop to a necessary standard.

It is critical to recognise that, when learning and applying STAMP / STPA, individuals are essentially rewiring their brains to think and act differently. This process needs time to develop and if individuals knowingly, or unknowingly, take a shortcut, they lose that opportunity to nurture and develop their skills.

When working through the analysis process, sudden realisations about potential causes of [hazards] and [losses] are surprisingly common, especially as analysts skill levels increase.

Unfortunately, whilst increased automation can help minimise repetition in the conduct of the analysis, it can also undermine the opportunity to experience those sudden realisations and undermine the creative exploration process shared amongst the analysis participants. This can lead to subtle but significant portion of [unsafe control actions], [hazards] and [losses] not being identified.

Analysts may miss a subtle issue on the first pass of a particular controller or control loop. However, as the analysis progresses, analyst skills increase, and other parts of the control structure are explored, logical links become apparent between those different parts of the analysis of the different parts of the control structure, and it is at this point that the subtle issue can become apparent. However, using increased automation may undermine that opportunity.

5.8 Time Spent is Time Lost: Use a Facilitator, Coach, Referee

Lastly, when starting out using STPA, and especially where the STPA analysis or its subject matter is not simple, it is highly recommended that support is gained, as early as possible, from people who can facilitate the conduct of the STPA analysis, coach the participants and referee where there are contentious findings.

This is to avoid spending valuable time trying to do it in isolation, going round in circles, second guessing and making mistakes. Time which could be better spent rapidly developing analysis skills and ultimately getting the STPA results (safety constraints) faster, and so start benefiting from the disproportionate positive impacts that early safety issue identification can have.

It's important to recognise that getting help sooner has a hugely disproportionate benefit in the medium to long term, especially where issues can be identified and resolved earlier in the engineering process. Furthermore, without support, there is also the possibility for missing things, or becoming distracted by less important issues, which could also have significant impacts on the project and ultimately safety.

As part of this project, having access to someone who the analysts could speak to and ask questions of was incredibly valuable for learning and making progress, as was the help to facilitate the analysis process. It also helped to minimise second guessing amongst the analysts and build their skills and confidence much more rapidly. Ultimately, this outside support helped to speed up the learning process and getting results.[5]

6 Conclusion

In conclusion, whilst analysing complex systems with STPA is no easy feat, especially when analysing an existing system, there are some benefits to the technique. Once it is acknowledged that learning and applying STPA is a challenging pathway and that initial challenge is overcome, the process becomes almost second nature and spotting (safety) problems early happens naturally, even without doing a fully detailed analysis. Of course, it's important to do a full analysis, but applying STPA in the early concept stages is hugely beneficial. STPA allows to foresee problems very quickly, possibly saving the industry millions in further development. Moreover, with the help of a coach, a facilitator to aid everyone, results occur more rapidly and you learn faster.

Acknowledgments The work of L. Buysse is funded by a "Travel Grant for a long stay abroad" by the "Research Foundation – Flanders" (FWO) (Grant no. V436222N). The research has received funding from VLAIO under grand agreement no. HBC.2020.2088 (Safety Assurance 4.0 -Management of Safety Risks in Industry 4.0).

References

Boehm, B. W. (1984). Software Engineering Economics. IEEE Transactions on Software Engineering, SE-10(1), 4–21. https://doi.org/10.1109/TSE.1984.5010193
Burton, S., Habli, I., Lawton, T., McDermid, J., Morgan, P., & Porter, Z. (2020). Mind the gaps: Assuring the safety of autonomous systems from an engineering, ethical, and legal perspective. Artificial Intelligence, 279, 103201. https://doi.org/10.1016/j.artint.2019.103201
Buysse, L., Vanoost, D., Vankeirsbilck, J., Boydens, J., & Pissoort, D. (2022). Case Study Analysis of STPA as Basis for Dynamic Safety Assurance of Autonomous Systems. In S. Marrone, M. De Sanctis, I. Kocsis, R. Adler, R. Hawkins, P. Schleiß, S. Marrone, R. Nardone, F. Flammini, & V. Vittorini (Eds.), Dependable Computing – EDCC 2022 Workshops (pp. 37–45). Springer International Publishing. https://doi.org/10.1007/978-3-031-16245-9_3
Dekker S. (2012). Just culture: balancing safety and accountability (2nd ed.). Ashgate.
Hentout, A., Aouache, M., Maoudj, A., & Akli, I. (2019). Human–robot interaction in industrial collaborative robotics: a literature review of the decade 2008–2017. Advanced Robotics, 33(15–16), 764–799. https://doi.org/10.1080/01691864.2019.1636714
International Organization for Standardization (ISO). (2010). ISO 12100:2010 : Safety of ma-

---

[5] People who help facilitate the analysis and coach the analysts do not necessarily need to know the detail of the subject matter. They should know just enough to be able to support the analysts / stakeholders, which is especially beneficial when applying STPA to sensitive / classified systems.

chinery - General principles for design - Risk assessment and risk reduction. https://www.iso.org/standard/51528.html#:~:text=ISO%2012100%3A2010%20specifies%20basic,designers%20in%20achieving%20this%20objective.

Leveson, N., Thomas, J. (2018). STPA Handbook. MIT. https://psas.scripts.mit.edu/home/get_file.php?name=STPA_handbook.pdf

PQRI – Manufacturing Technology Committee – Risk Management Working Group. (2015). Risk Management Training Guides: Hazard & Operability Analysis (HAZOP). https://pqri.org/wp-content/uploads/2015/08/pdf/HAZOP_Training_Guide.pdf Accessed 25 October 2022

Siemens. (2016). How to conduct a failure modes and effects analysis (FMEA). https://polarion.plm.automation.siemens.com/hubfs/Docs/Guides_and_Manuals/Siemens-PLM-Polarion-How-to-conduct-a-failure-modes-and-effects-analysis-FMEA-wp-60071-A3.pdf Accessed 25 October 2022

Whiteley, S. (2022a). STPA Hazard Analysis Foundation Course v1.0. https://www.whiteley-safety.co.uk/210309-stpa-lp Accessed 25 October 2022

Whiteley, S. (2022b). STPA Hazard Analysis Tables. https://www.whiteley-safety.co.uk/ Accessed 25 October 2022

Zhang, T., Li, Q., Zhang, C., Liang, H., Li, P., Wang, T., Li, S., Zhu, Y., & Wu, C. (2017). Current trends in the development of intelligent unmanned autonomous systems. Frontiers of Information Technology & Electronic Engineering, 18(1), 68–85. https://doi.org/10.1631/FITEE.1601650

Build, monitor, and measure your live safety case in nLoop

Carmen Cârlan

Edge Case Research GmbH

Abstract *To help AV companies to deploy safe, trustworthy autonomy, Edge Case Research proposes a new product – nLoop, which supports a new working model for building autonomous systems where the entire organization speaks the language of safety and measures progress continuously. nLoop's live safety cases, requirements tracing, hazard tracking, and test coordination help teams achieve the goal of building AVs that are safe enough to deploy. At its core, nLoop supports the specification and management of structured safety cases. The validity status of the claims within nLoop safety cases may be evaluated by the evaluation of defined Safety Performance Indicators (SPIs), which, according to UL 4600, are metrics for assessing the system's safety performance. SPIs in nLoop are continuously evaluated based on the data coming from safety evidence providers (e.g., databases, issue tracking systems, external verification and validation tools) connected to the safety case via dynamic links, where the dynamic links allow the safety case to be aware of newly generated evidence both during design- and run-time. The evaluation of the SPIs enables the evaluation of the claims within the safety case, and, implicitly, the evaluation of the entire safety case. Consequently, safety cases in nLoop are 'live', keeping track of the status of the current safety performance of the overall system, given the available safety-relevant evidence. The validity status of the claims in the safety case may be used as feedback for system developers. Given the invalidation of a safety claim, system developers and operators may update the system so that the desired system safety performance is (re)established. System updates usually imply the generation of new safety evidence, which, via safety evidence providers, is used again for evaluating SPIs, thus creating a feedback loop between the safety case and the activities executed by system developers and operators.*

© Carmen Cârlan 2023.
Published by the Safety-Critical Systems Club. All Rights Reserved.

A Service Analysis of the Mont Blanc Tunnel Fire

James Catmur, J C and Associates

Kevin King, BAE Systems

Mike Parsons, AAIP, University of York

Fathi Tarada, Mosen

Abstract   On 24th March 1999 a transport truck caught fire while driving through the Mont Blanc tunnel between Italy and France. Other vehicles travelling through the tunnel became trapped and fire crews were unable to reach the transport truck. The fire burned for 53 hours and reached temperatures of 1,000°C producing toxic smoke. Authorities compounded the problem by pumping air from the Italian side, feeding the fire and forcing poisonous black smoke through the length of the tunnel. A total of 39 people were killed. In the aftermath, major changes were made to the tunnel to improve its safety. This paper analyses the accident from a Services perspective, examining which critical services were in use (fire alert service, tunnel control service, ventilation service, etc), and which contributed in some way . The Safety Critical Systems Club (SCSC) Service Assurance Guidance v3.0 is used to guide the analysis and provide structure to the work, producing a service hierarchy map, criticality levels and identification of assurance needs. The improvements put in place after the accident are assessed to see the effect on the service assurance involved.

1 Introduction

This paper aims to show how a paradigm of Service Assurance combined with a suitable stepwise methodology enables analysis of organisational structures that contributed to a severe accident in the Mont Blanc tunnel. It also allows the service improvements made after the accident to be evaluated to see how they affect the overall assurance position.

A road tunnel is a good example to apply the technique of Service Assurance to because a tunnel provides a pure service (in this case the service of point-to-point travel) to a range of road users with no tangible product being delivered.

© James Catmur, Kevin King, Mike Parsons, Fathi Tarada 2023.
Published by the Safety-Critical Systems Club. All Rights Reserved.

1.1 Service Assurance

Many current safety systems rely on functionality provided by services which are designed, developed, operated, and maintained outside the immediate boundaries of the system. In many cases, overall system design is essentially about managing the interactions between various service functionalities which co-operate to produce a useful effect in an operational scenario.

This approach is highly applicable to the idea of a *tunnels-service*, i.e., the provision of a suitable through-route for road vehicles to run on, the signage for the driver, and supporting maintenance and development of the tunnel infrastructure. All the services contributing to the overall *tunnels-service* need to be provided to the appropriate level of quality and safety, and tangible assurance artefacts are needed to show this is the case.

This work looks at how such a services view of tunnels can be applied in a retrospective study of the Mont Blanc Tunnel Fire accident and shows how this can lead to a useful assurance framework that highlights not only the contributory factors to the accident but also subsequent improvements to the tunnel systems and infrastructure to improve the overall *tunnels-service*.

In recent years there has been increasing interest in the topic of safety-related services. There is a range of literature produced over the years looking at assurance of services. Some of this has been produced by the SCSC Service Assurance Working Group (SAWG) itself and has been presented at conferences such as the SCSC Safety-Critical Systems Symposium (Catmur et al 2022, Durston et al 2019, Elliot and King 2019, Harris et al 2019, King et al 2020). The SAWG also produces a guidance document on service assurance which has undergone evolution over the years (SAWG 2020, SAWG 2021, SAWG 2022). There is also an interesting paper considering a service perspective for education, research, business and government, produced by the University of Cambridge Institute for Manufacturing (IfM) and International Business Machines Corporation (IBM) (IfM and IBM, 2008).

1.2 Rationale

The main reasons why this approach is useful are:
1. Tunnels are operated and maintained as a service provision for the user.
2. A tunnel does not produce or modify anything material for its user.
3. A service-based approach to assuring safety provides a different, useful, and important perspective.
4. A service-based approach to safety includes the impact of organisations, agreements, and contracts. It is the only safety assurance approach to do so.

5. It is recognised that collaborative working of technology, organisations, people, and processes all contribute to safety and need to be part of the picture.
6. A service approach recognises the concept of time-limited contracts which are appropriate for road operation and maintenance.
7. There is a significant shift to a service-based approach in many areas of technology and commerce, and hence it is worth exploring an established service delivery example.

1.3 Limitations of study

We aim to provide an illustrative example as some contract information is no longer available or commercially sensitive and not available in the public domain. Where necessary we have proposed some of the missing details. This does not detract from the application of the service assurance methodology.

2 Introduction to Service Assurance

Service Assurance is a different way of producing trust or confidence in something which is not product-based; it is based on information about organisations, contracts and other information relating to the delivery of the service. For this reason, it is a much more appropriate method of gaining confidence in something where there may be no tangible deliverables. Tunnels (strictly the journey through the tunnel) is such an example, where we need to have confidence that it can be undertaken safely but nothing material is "handed over" to the person making that journey in a vehicle.

2.1 Definitions

Some service-based definitions are given below:

Term	Definition
Operational Level Agreement (OLA)	Defines the interdependent relationships in support of a Service Level Agreement (SLA). The agreement describes the responsibilities of each internal group toward other groups, including the process and timeframe for delivery of their services. The objective of the OLA is to present a clear, concise and measurable description of the service provider's internal support relationships.

Term	Definition
Service-Based Solution (SBS)	An SBS comprises the systems, organisations, processes, and resources to deliver and manage the services through the duration of the contract life. It may consume other services.
Service Catalogue	A Service Catalogue is the commercial document that lists and describes the services offered for consumption. It is constructed by the Service Provider and typically does not give any service implementation details.
Service Consumer	A Service Consumer consumes (i.e., makes use of) one or more Services
Service Contract	A Service Contract is the legal agreement between Service Provider and Service Consumer. Note that the Service Consumer may not be involved in defining the service or the SLAs at the outset; they may be provided, pre-defined and pre-packaged by the Service Provider on a take-it-or-leave-it basis
Service Definition	The Service Definition describes the services available for consumption which may include technical and/or commercial aspects. It may include deliverables, prices, contact points, availability, ordering, and processes to request Services. This may include a Service Catalogue.
Service Level Agreement (SLA)	An SLA is the agreement between the Service Provider and Service Consumer that defines the level of service (e.g., in terms of availability, performance, and quality) that the Service Consumer will receive. It often has targets for each service described in the Service Catalogue. It usually specifies responsibilities of the Service Provider and Service Consumer and defines the penalties in the event that the specific targets in the SLA are not met.
Service Provider	A Service Provider provides (i.e., offers to consumers) one or more Services.

2.2 What is a Service?

The way that a Service is normally described or defined is different from the specifications and descriptions more commonly used in safety-related systems. An individual Service (sometimes called a *Service Component*) is typically offered by a *Service Provider* via an entry in a *Service Catalogue*. The Service Catalogue usually describes the capabilities/functionality offered to a *Service Consumer* without providing much (or indeed, any) of the implementation detail, in fact it is unusual for the design and implementation of the Service to be visible to the Consumer. Note that service catalogue is a commercial document as well as a technical one and it may give information such as the hours a service is available, level of support, etc.

A Service Level Agreement (SLA) is used to define the level of service being offered, this may include functional and non-functional properties such as capacity, performance, and availability. SLAs often describe (commercial) penalties on the Service Provider for not meeting key elements of the agreements. Typically, penalties are framed in terms of service credits, but may be also in different terms.

Service Contracts between the Provider and Consumer provide the overriding legal and commercial picture and typically refer to Service Catalogues, Statements of Work (SoW) and SLAs.

The boundary between a Service Consumer and a Service Provider is typically both an organisational and commercial boundary as well as a technical one. A Consumer may not be involved in the specification and development of a Service and instead may select a commodity or standardised Service (i.e., something already widely available). Alternatively, they may be involved in the creation of new, tailored, or bespoke services.

2.3 Service Context and Service-Oriented Architecture

Figure 1 gives the context of Service Provision and Consumption:

Fig. 1. Context of Service Providers and Consumers (SAWG 2022)

2.4 Simplified View of Tunnel Services

This paper is concerned with the service analysis of the Mont Blanc Tunnel. This is necessarily a simplified view but serves as a very useful and real example of how the service assurance approach may be applied.

Note that services are used extensively in a highways network, not only for provision and maintenance of the highway, but also for areas such as breakdown and accident management.

2.5 The Service Assurance Guidance

The Service Assurance Guidance Document v3.0 (SAWG 2022) provides a framework for safety assessment of services, together with principles and objectives for assuring them. The main elements are:

The Introduction explains why services in a safety context are problematic. It covers background aims and scope, and the target audience. The overall approach is that the document is positioned as guidance; it may be used for developing (domain-specific) standards and further guidance for services. It discusses views of what a service is and what service characteristics are. It also introduces service terms used in the document.

The Assurance of Services section begins by introducing some of the challenges of assuring services to describe what is different about services (as opposed to systems) from an assurance view. It introduces further concepts and terms relevant to assurance of services. Finally, it lists some basic assumptions.

The key part of the document, Service Assurance Principles, states the *Six Service Assurance Principles*, including brief supporting descriptions and explanations. These are:

1. Service assurance requirements shall be defined to address the service-based solution's contribution to both desirable and undesirable behaviours
2. The intent of the service assurance requirements shall be maintained through the service definitions, service levels, the service architecture and the agreements made at service interfaces
3. Service assurance requirements shall be satisfied
4. Unintended behaviours of the service-based solution shall be identified, assessed and managed
5. The confidence established in addressing these principles shall be commensurate with the level of risk posed by the service-based solution
6. These principles shall be established and maintained throughout the lifetime of the service-based solution, resilient to all changes and re-purposing

It then defines objectives which support each principle; these are seen as a route of demonstrably meeting the principles. There is also a mapping of the principles to service characteristics.

The concept of Levels of Service Assurance (LSA) is described next. The levels are then used to scope the applicability of objectives, so tailoring what is required for each level of service risk.

The Capturing Justifications and Evidence section provides evidence tables covering aspects of service scoping, design, analysis, implementation, and change. These tables suggest evidence techniques and containers for meeting the objectives. The concept of Assurance Wrappers is introduced and explained. Some further service assurance challenges and some solutions are discussed.

A brief discussion of possible assurance techniques is given in the Analysis Techniques section with the most promising techniques identified for further work.

The document also provides extensive supporting sections including the following topics: (i) Service 'Mode' Changes, (ii) What Happens when Services Go Wrong? (iii) Further work, (iv) a set of 'Hazop'-style guidewords for services,

(v) a set of service-related Incidents and Accidents as identified from publicly available sources and (vi) a workflow for analysing services.

3 The Fire

3.1 Technical Characteristics

At 11.6 km long, the Mont Blanc tunnel is one of the longest road tunnels in the world with two-thirds (7.64 km) lying in French territory and one-third (3.96 km) in Italian territory. Although a tunnel under Mont Blanc was first considered in the 19th century, the idea only gained real attention in the early 1900s when preliminary designs were presented to politicians from both France and Italy. However, the political turmoil across Europe that led to World War I and subsequently World War II delayed re-consideration of a tunnel until the late 1940s, with drilling eventually beginning in 1959. By August 1962 the teams drilling from each side met making the tunnel a reality, with it finally opening to traffic in July 1965.

At the time of opening the Mont Blanc Tunnel was three times longer than any other road tunnel across the world, earning it the nickname "The World's Longest Shortcut" (Figure 2). It also reduced the journey distance between Paris and Rome by 150km.

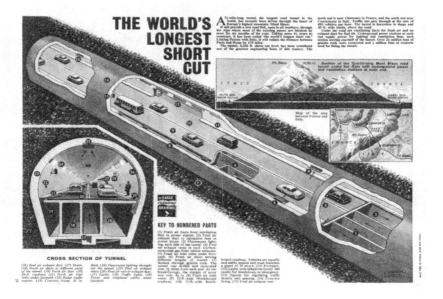

Fig. 2. "The World's Longest Short Cut"

Compared to similar two-way road tunnels the Mont Blanc tunnel is relatively narrow at 7m (Figure 2, bottom left); some recent two-way tunnels are as wide as 9m.

It is also important to highlight that the tunnel is at a high altitude compared with others, being 1,274 m above sea level at the French entrance and 1,381 m at the Italian entrance. This is relevant because heavy goods traffic reaches the tunnel entrances after having climbed long and quite steep roads to get there.

The tunnel consists of a single bore running through mountains which are more than 2,000 m thick above the tunnel for more than half of its length.

Similar tunnels such as at Saint-Gothard and Fréjus have shafts intermittently positioned to provide ventilation. Mont Blanc does not, as this was considered impossible, meaning that to ensure adequate ventilation to dilute exhaust gases, fresh air could only be taken into the tunnel from either end, circulating through shafts under the roadway and distributed into the main tunnel through air vents spaced every 10m and located at the level of the road surface on one side of the road (Figure 3).

A separate duct located under the road surface is used to extract polluted air through vents positioned every 300m along the tunnel.

Several features were installed into the tunnel as part of its original design to enable safe travel. Lighting, emergency telephones, and refuges (known as *garages*, 3.15 x 30 m long with turning places) located every 300m, that alternate between the two sides of the road. There were also traffic signals positioned every 1.2km through the tunnel.

Fig. 3. Tunnel construction showing ducting under the roadway (GEIE-TMB, 2022)

The two concession companies made significant changes to the tunnel facilities since it first opened, primarily to aid safety and comfort but also to increase capacity. In 1979, due to the increase in heavy goods traffic the ventilation system was adapted with the duct, originally used to extract polluted air, repurposed to be reversible, allowing for the introduction of extra fresh air into the tunnel. Further improvements included systems installed to selectively concentrate fume extraction where needed most; shelters installed every 600m; pressurised water systems for firefighting and the complete replacement of the lighting systems.

By March 1999, studies were underway or in planning to consider systematic automated management and automatic detection of incidents.

It is important to point out that although other bi-directional road tunnels around the world may have lower volumes of traffic (although not all), they were generally no better equipped from a safety perspective at the time of the incident.

3.2 Operational Characteristics

To manage the day-to-day operations of the tunnel from the outset it was agreed that two separate concessions would be created, Société du Tunnel du Mont-Blanc (STMB), which later became the Autoroutes et Tunnel du Mont-Blanc (ATMB), under French jurisdiction and the Società Italiana per il Traforo del Monte Bianco (SITMB) under Italian jurisdiction. It was agreed that although the territorial split was 65% French, 35% Italian, each concession would be accountable for half of the tunnel.

Since its first opening, there had been a marked increase in traffic flow. By 1999 small vehicle traffic had more than doubled from the original estimated usage, which although significant, is modest compared to other tunnels today. Heavy goods vehicle traffic however had increased more appreciably, by a factor of 17 over 34 years, showing the relative importance of the tunnel for trade between the two countries. In 1998 the tunnel was projected to take c.960,000 heavy goods vehicles, however in reality it accommodated almost 2 million (c. 5,600 vehicles per day).

The traffic regulations in place at the time stated that moving vehicles must be at least 100 m apart and restricted to 50-70km/h. However, there was no provision for the distances between stationary vehicles. Dynamic signage was originally placed in the tunnel but was subsequently removed by the two concessions.

In the tunnel, there were traffic lights every 1200 m. During the fire these lights were switched to red a few minutes after the alarm was sounded, but they did not reduce the toll of the disaster, either because some of the lights in the tunnel did not work or because they were not respected (the lights were not very visible).

In 1995, radio broadcasts were installed in the tunnel. First, to transmit communications on service and safety frequencies; then to allow messages to be broadcast to users on two French and two Italian public frequencies. The radio broadcasts were not used to communicate with users after the fire alert.

3.3 Concerns around the Tunnel Operation Prior to the Fire

The initial joint French-Italian enquiry report (Duffe et al, 1999) picked up several concerns relating to the operation of the tunnel before the fire. There were two separate command and control stations, one for each half of the tunnel operated by its respective concession; the tunnel safety regulations were partially obsolete; the smoke extraction capacity was half that of comparable tunnels; there was no independent safety passageway to facilitate the transit of emergency services or the evacuation of users, and the rescue services were organised differently on each side of the tunnel.

Furthermore, the separate concessions were not made aware of the existence of a French regulation from 1981 which defined the safety conditions to be applied in road tunnels, particularly that safety installations in older tunnels should keep pace with the evolution of traffic. This was because the bi-national agreements governing the operation of the tunnel did not refer to any specific safety regulations.

It is also important to understand that before the fire the two concessions often found it difficult to coordinate investments in tunnel infrastructure. The studies underway were agreed upon jointly and concerned improvements to the whole tunnel but these were only in planning when the fire occurred.

3.4 The Timeline

On the morning of 24 March 1999, after a relatively unblemished operating record, a Belgian heavy goods vehicle with a refrigerated trailer made from combustible isothermal foam (here after called 'the truck') entered the tunnel travelling South-East from France to Italy. It was carrying 12 tonnes of flour, 9 tonnes of margarine, and 550 litres of diesel (Bailey, 2004)

The events of the incident from the time at which the truck entered the tunnel are summarised in Table 1.

Table 1. Summary of Events

Time (C.E.T.)	Event
10:46 Wed 24th	The truck entered the tunnel from the French side
10:52	Oncoming drivers noticed smoke coming from the truck and began flashing their headlights at the driver. By this time the truck was around 3km into the tunnel. An obscuration detector was activated and the operator's screens at the French control station showed smoke in the tunnel.

Time (C.E.T.)	Event
10:53	Noticing the smoke coming from underneath the truck's cabin the driver pulled over and attempted to fight the fire himself. At this point the incident was not considered a fire emergency as there had been 16 other truck fires in the tunnel since it opened, in each case extinguished by the driver. The difference on this occasion was the driver being forced back when the truck burst into flames, forcing him to flee the scene through the Italian entrance to the tunnel. The nearest sensor of the fire detection system, operated by the Italian concession, had been taken out of service the night before.
10:54	A driver from one of the nearby vehicles who had escaped to refuge 22, raised the alarm.
10:55	The fire alarm was triggered by tunnel employees who stopped further traffic from entering. By this time a further 10 cars and 18 trucks had already entered the tunnel from the French side. Some of these cars managed to turn around in an attempt to retreat back to French territory. However, air was flowing through the tunnel from the Italian side rapidly forcing dense smoke from the fire down the tunnel and quickly making navigating the road impossible. Turning around was not an option for the trucks and neither was reversing out of the tunnel.
11:00	Without oxygen from clean air, car engines began to stall leaving 50 people trapped in vehicles close to the fire. Most rolled up their windows and waited for rescue, but some escaped, and where not overcome by the heat or toxic components of the smoke (mainly cyanide), sought refuge in the fireproof shelters built into the walls of the tunnel. Fire crews responded – two French fire trucks from Chamonix and an Italian one from Courmayeur.
11:11	A second crew of Italian firefighters arrived at their end of the tunnel.
11:15	The first French fire truck managed to reach a point 2.7 km from the truck on fire, however a combination of failed lighting and abandoned vehicles made it impossible for them to get closer to the truck.
11:20	The first Italian fire crew managed to get within 300 metres of the truck before being forced to abandon their vehicles and take shelter in one of the fireproof refuges, 0.9 km from the truck. As they took refuge the fire began to spread with burning fuel flowing down the road surface, causing tyres and fuel tanks on abandoned vehicles to explode, and sending deadly shrapnel in the air.

Time (C.E.T.)	Event
11:25	The second Italian fire crew had to abandon their vehicles, ending up searching on foot for their trapped colleagues. Realising the cubicles were offering little protection from the smoke, they began searching for the doors that led to the ventilation duct.
11:30	The second French crew could only reach 4.8 km from the truck before they were stopped by the heat and smoke. By this time, smoke had reached the French entrance to the tunnel, 6 km from the truck.
11:54	A rescue mission was instigated using the fresh air channels located under the road, saving the lives of some people, however that took over 6 hours to enact.
13:04	French specialised emergency plan was activated for the tunnel.
13:35	A separate 'Red' plan was activated following the blockage of the rescue workers.
18:35	The civilians who had managed to escape their vehicles to the relative safety of the refuges were sadly not out of danger as the fire doors on the cubicles were only rated to survive for two hours and it took over 7 hours to reach those people. 6 civilians were eventually saved from refuge 17.
16:00 Friday 26th	The fire is finally brought under control

The fire burned for 53 hours and reached temperatures of 1000°C with the extreme temperatures mainly due to the margarine on the trailer. It has been assessed that the trailer was equivalent to a 23,000-litre oil tanker. The fire spread to other cargo vehicles nearby that also carried combustible materials (Figure 4).

Fig. 4. Wreckage inside the Mont Blanc Tunnel (Bettelini, 2022)

Thankfully, the heroic efforts of the firefighting teams and a security guard, who perished in the fire after evacuating 10 survivors on his motorcycle, 12 of the 50 people trapped in the tunnel survived being evacuated to the Italian side. All 15 of the trapped Italian firefighters were also eventually rescued, 14 of them recovered but sadly their commanding officer died in hospital. Investigation discovered that most of the victims died within 15 minutes of the fire first being detected.

Due to weather conditions at the time, airflow through the tunnel was from the Italian side to the French side (Bailey, 2004). The shape of the tunnel led to it acting like a chimney, an effect compounded by the authorities pumping in further fresh air from the Italian side, feeding the fire and forcing poisonous black smoke through the length of the tunnel. Only vehicles closer to the entrance on the French side of the tunnel were trapped, while cars on the Italian side of the fire were mostly unaffected.

The intensity of the heat was evident as after extinguishing the fire it took a further five days for temperatures to cool down enough for debris to be removed from the tunnel.

3.5 The Aftermath

The tunnel underwent major enhancements before it re-opened in 2002 (Bailey, 2004). Improvements include computerised detection equipment, extra security bays, a separate, sub-road surface, escape tunnel and a fire station in the middle of the tunnel. The safety tunnels have clean air flowing through them via vents, separate to the main tunnel. The security bays now have direct video contact with the control centre, so they can be informed about what is happening in the tunnel more clearly.

Remote sites for safety inspection of all heavy goods vehicles were also created on each side of the tunnel way before the tunnel entrance. These remote sites are also used as staging areas, to smooth the flow of commercial vehicles.

Most importantly the management of all services relating to tunnel maintenance and operations were subsumed under a single operating authority that was jointly owned by the French and Italian authorities, The Gruppo Europeo di Interesse Economico del Traforo del Monte Bianco (GEIE-TMB).

4 Service Analysis Pre-Fire

4.1 Services Identified

For this paper, it was important to identify the services in place for the Mont Blanc tunnel at the time of the fire. Specifically, those services whose fallibility may have contributed to the incident.

Evidence of the services in place from when the tunnel opened in 1965 through to the time of the tunnel fire in 1999 is limited. There are a handful of academic papers and articles that consider aspects of the tunnel operation in relation to the fire that were published after the incident (e.g., Bettelini 2022, Bailey 2004, Lee & Ghazali 2018) and of course there is the official accident report (Duffe et al 1999).

To operate any road tunnel the consensus (e.g., Bettelini 2022, Lee & Ghazali 2018) is that tunnels need signage, lighting and ventilation, all managed by control staff in some form of control facility. Importantly, considering that accidents do happen, it is also crucial to have an effective, maintained firefighting capability. Each of these aspects can be considered as a service and should have adequate assurance.

That consensus enabled us to hypothesise that the service structure at the time of the accident would most likely have been analogous to the hierarchy set out in Figure 5[1], with separate service structures in the two halves of the tunnel: French and Italian. We have shown the possible structure of the services for the French (ATMB) half of the tunnel as our hypothesis is that the Italian service structure would have been analogous to the French, so to ensure the hierarchy is legible we have left the Italian half un-developed.

---

[1] Given the elapsed time since the fire we cannot be certain how these services were structured, but the available material suggests a service structure akin to this.

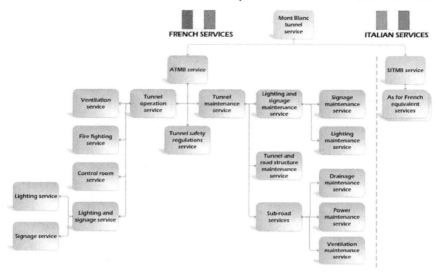

Fig. 5. Mont Blanc Tunnel Service Hierarchy

4.2 Applying the Service Assurance Guidance

The Service Assurance Guidance v3.0 (SAWG 2022) provides a structure that enables services to be assured against each of the *Six Service Assurance Principles* through approaches aligned to the *Service Objectives* that underpin each principle. The structure enables us to consider services from several perspectives; how they inter-relate, in what could be considered a service stack or hierarchy, the level of safety risk inherent in the use of a service and how much assurance is offered by each service provider or needs to be developed by the relevant service consumer to address assurance shortfalls.

Key for our analysis of the services supporting operation and maintenance of the Mont Blanc *tunnels-service* before the fire was our understanding of the service hierarchy (Figure 5), an assessment of the Levels of Service Assurance and awareness of the quality of assurance evidence in place for each service in the hierarchy.

4.2.1 Levels of Service Assurance

The term Level of Service Assurance (LSA) has been developed by the SAWG as it is understood that, in a services context, the type of service and manner in which it is consumed will require differing levels of assurance. Classifying[2] a service with a LSA formalises this approach, so it becomes explicit as to what is required to meet a particular level of safety risk inherent in a service. Five levels of LSA have been developed (Table 2), the higher the level the more objectivity you need to assure the associated service.

Table 2. Levels of Service Assurance (SAWG, 2022)

Level of Service Assurance	Definition (Service Consumer View)
LSA 0	No safety aspects present in the service, so no objectives assigned
LSA 1	Minor safety aspects with little impact of failures (minor injury possible but unlikely)
LSA 2	Safety aspects with some impact of failures (several injuries possible)
LSA 3	Significant safety aspects with service with major impact (could indirectly lead to multiple injuries or a single death)
LSA 4	Service is safety-critical[3]: service failures could have catastrophic impact (could directly lead to multiple deaths)

LSAs are allocated in a top-down fashion, starting with the highest-level service element in a hierarchy (i.e., the service visible to the consumer) and then flowing down to sub-ordinate services. Importantly, an allocation rule is applied, whereby at least one of the subsidiary services inherits the LSA from its parent service[4].

We postulate that applying the LSA rules from the Service Assurance Guidance to each service in our assumed hierarchy for the Mont Blanc tunnels-service (Figure 5) gives us the LSAs overlayed in Figure 6.

This is a proposal of the LSA with respect to the tunnel user, and we suggest operations services have a higher LSA than maintenance ones, as they are closer to the user. LSAs for staff including emergency services personnel may also have a different balance.

---

[2] It is important to note that the LSA is determined first by the service consumer; the service producer then shows how the service meets this.

[3] It is acknowledged that the term 'safety-critical' as related to services is not a formally defined term. We suggest 'a service whose failure may result in one or more of the following: death or serious injury to people; loss or severe damage to equipment or property, or widespread environmental harm'

[4] So, for example if the top-level service is LSA3, then at least one of the next-level down services in the hierarchy is also LSA3 (or possibly LSA4).

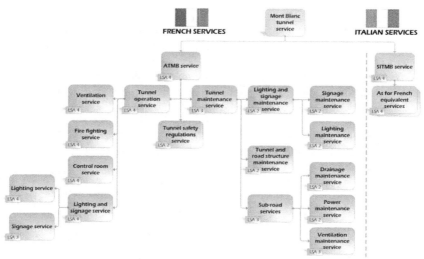

Fig. 6. Mont Blanc Tunnel Service Hierarchy with assessed Levels of Service Assurance

4.2.2 Completeness of Service Assurance Evidence

The Service Assurance Guidance v3.0 (SAWG 2022) has established the concept of *service assurance wrappers*, as a mechanism to bridge the gap between what is required to support achievement of an LSA and the actual evidence available. Assurance across the service boundary needs to meet the assessed LSA, but it is unrealistic to expect that assurance to be provided wholly by the service provider. *Service assurance wrappers* need to cater for differing scenarios.

This has led to service assurance wrappers being expressed as having a key characteristic of 'thickness'. The spread of service assurance between the provider and the consumer informs the 'thickness' of the wrapper[5].

Wrapper complexity is, however, not a simple function of the class of wrapper as it will also be influenced by the degree of dependence on the consumed service and the nature of its integration into the overall hierarchy of services.

Applying the service assurance wrappers approach to our hypothesis of the service hierarchy for the Mont Blanc *tunnels-service* we were able to specify the

---

[5] Full thickness (class 3) wrappers address the situation where assurance requirements are not flowed down to the provider and minimal (or no) assurance is supplied by the provider. Medium thickness (class 2) wrappers need to map the assurance evidence from the provider to the context and level of risk the service is exposed to by the consumer. Thin (class 1) wrappers are applied where the service provider is a capable, 'safety-aware' supplier that either fully understands the consumer's domain and the use of their service within it (Class 1a) or at worst is still able to provide a safety-assured service even if they are unaware of the consumer's domain (Class 1b).

applicable types of *wrappers* we would have expected and their respective thicknesses (Figure 7). We will consider later whether such wrappers existed and their veracity.

It should be noted that an organisation that develops an assurance wrapper may be the service consumer itself, the service provider, or a third party. However, the consumer of a service remains accountable for the claims made in the associated assurance wrapper. At the highest level of the architecture, where safety risk can be understood, this will be the 'Duty Holder' for the overall system. Before the tunnel fire there was no obvious single 'Duty Holder' in this context, rather ATMB and SITMB were acting independently with no overarching body co-ordinating the two organisations.

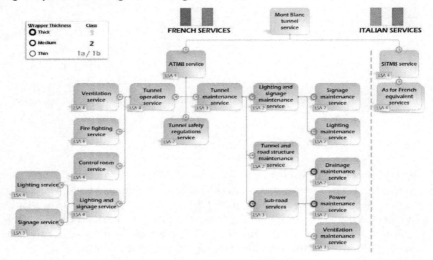

Fig. 7. Mont Blanc Tunnel Service Hierarchy with assessed LSA and Service Wrappers

4.3 Analysis in the Context of the Fire

Considering the available evidence (Duffe et al 1999, Bailey 2004), we believe elements of the history and characteristics of the tunnel (see Section 3.1 and 3.2) were important factors in how the 1999 fire incident evolved, e.g., the lack of ventilation shafts. Examining the accident timeline through the lens of service assurance we were able to determine which services may not have been fully assured and, importantly, the impact from the lack of coordinated services between the two halves of the tunnel.

From our hypothesis of the service hierarchy, LSA and Service wrappers (Figures 5, 6 & 7) we have been able to analyse the assurance deficits in the context of the fire. As we believe the service hierarchies on each side of the tunnel were analogous, we have focused our analysis on the services we believe were provided by ATMB for the French 'half' of the tunnel.

The overall service 'package' provided by ATMB is assessed to be at LSA4. If we follow the SAWG Guidance (SAWG 2022) it is correct to analyse that service first. However, it is broadly split into three areas of supporting services[6] where we believe the assurance evidence should sit, *tunnel operation services*, *tunnel maintenance services* and a *tunnel safety regulations service*; so, it is sensible to focus on these services (in LSA order) to understand the areas of deficiency in the ATMB service.

4.3.1   ATMB tunnel operation service (LSA4)

Our hierarchy (Figures 5, 6 & 7) breaks down the tunnel operation into four subordinate areas of service provision, a *control room service*, *lighting and signage service*, *firefighting service,* and *ventilation service*. We believe the reliance on these services requires them all to be assured to the highest level, LSA4. This directs to them needing to substantively meet the objectives of each of the *Six Service Assurance Principles*. Therefore, it is appropriate to expect, as a minimum, the evidence listed in Table 3 would be available from the service provider, or else need to be developed as part of an assurance wrapper (see section 4.2.2).

Assessing the four subordinate services against those evidence requirements enables us to highlight the areas of weakness in service assurance (see Table 3).

The most obvious weakness in respect of the ATMB service provision is that it stopped halfway through the tunnel. There were two separate command and control stations, one for each half of the tunnel operated by its respective concession with the *control room services* delivered by those stations not synchronised. On top of this lack of a co-ordinated service interface we can find no evidence of a common public emergency plan, with the rescue services organised differently on each side of the tunnel, working to separate emergency response plans, evident by the response times and actions as discussed in section 3.4.

The Italian authorities did not activate their specialised rescue plan, believing the accident had occurred on French territory. Meanwhile the French had two plans, their specialised emergency plan for the tunnel and a red plan to manage the rescue services. These plans required command posts, notably an operational command post (PCO) at the ATMB site. Sadly, the PCO did not have means of autonomous, direct communication with their Italian counterparts.

---

[6] Note that interactions between services are covered by Principle 4 - Unintended behaviours of the service-based solution shall be identified, assessed and managed: All undesired or unintended behaviours which may impact safety properties or safe behaviour of the overall system must be identified and assessed within the usage context.

Table 3. Evidence for assurance of the ATMB tunnel operation services

Expected Evidence	Control room service (LSA4)	Lighting and signage service (LSA4)	Fire-fighting service (LSA4)	Ventilation service (LSA4)
✓ = evidence available OR service likely compliant, even if no clear evidence ? = no evidence in public domain and unclear if service was compliant ✗ = service weakness or failure may have contributed to the fire				
An understanding of the context and intended use of the Service by all stakeholders	✓	✓	✓	✓
Clear understanding of the requirements, stakeholders and consumers of the Service	✗	✓	✓	✗
All the states or modes of operation of the Service, including degraded modes are clearly defined	✗	✓	✓	✗
Full awareness of any necessary industry standards, legislation or regulation applicable to the design of the Service, including where applicable reference to similar existing designs	✗	?	✓	✗
Relevant redundancy, standby modes and diversity in the Services design is documented and understood	✓	✓	✓	✓
A clear breakdown of the components of the Service, their interaction and their interfacing to other services or products is documented and understood	?	?	✗	✗
Documented understanding of the risks and hazards presented by the Service, including any controls or mitigations in place to manage those risks to an acceptable level, e.g. So Far As Is Reasonably Practicable	✗	✓	✗	✗
How the assurances that the services are acceptably safe are routinely tested, e.g. by independent audit	?	?	?	?
How the Services are staffed, and the quality of the staff	?	?	?	?

The poor coordination between the two concessions appears to have been true of many services, with the *ventilation service* and *firefighting service* each being key on the day of the accident: and varying investment in services over the long-term resulting in a 'tunnel of two different halves'.

Looking at some of the key operational concerns predicted before the fire, it is not clear from the evidence we have seen that ATMB understood the breakdown of the services they were accountable for and the need for those services to evolve to keep pace with the evolution of traffic. At the time of the fire the smoke extraction capacity was half that of comparable tunnels which, coupled with the lack of co-ordination, led to the Italian's *ventilation service* pumping extra clean air into the tunnel unaware that it was simply providing more oxygen to feed the escalating fire and causing additional smoke logging of the tunnel.

The development of the fire and significant issues experienced by the rescue services show that the design of the tunnel had not taken sufficient account of the need to facilitate the transit of emergency services to the site of a fire or the evacuation of users from it. It is our opinion that the *firefighting service* could only have worked if there was a far more adequate ventilation system to control the noxious smoke coupled with an independent safety passageway to facilitate fast transit to/from the scene.

4.3.2   ATMB tunnel maintenance service (LSA3)

As with the ATMB tunnel operation service, our hierarchy (Figures 5, 6 & 7) breaks down the tunnel maintenance into subordinate areas of service provision. We see there being three, *Sub-Road services* (LSA3) - including drainage, ventilation, and power, *Tunnel and road structure maintenance service* (LSA2) and *Lighting and signage maintenance service* (LSA2). Although we see these services requiring lower levels of assurance it is still appropriate that they comply with the objectives of each of the *Six Service Assurance Principles*. We believe, as a minimum, the same evidence listed in Table 3 for the operational services would be necessary and available either from the service provider, or else needed to be developed as part of an assurance wrapper (see section 4.2.2).

We were unable to determine if any of the *maintenance services* were fallible in a way that may have contributed to the fire. The *maintenance services* were probably delivered in a way that supported the respective operational service and it was those operational services where the concerns lay over their fallibility.

At the time of the fire, the firefighting service was managed independently of the tunnel by the local fire services in Chamonix (French) and Courmayeur (Italian) so there is no *firefighting maintenance service* in scope of our assessment.

4.3.3 ATMB tunnel safety regulations service (LSA2)

If operational oversight took account of existing legislation when the tunnel opened to the public and kept pace with such legislation as it evolved, then a thin wrapper of an assurance case owned by the operating authorities on each side of the tunnel would seem appropriate. We were unable to find any supportive evidence in the public domain regarding the assurance of the *tunnel safety regulations service* before the fire. Therefore, we cannot comment as to whether legislative compliance was being maintained by ATMB. We did however find that ATMB do not appear to have picked up a 1981 change in French legislation requiring tunnel safety to keep pace with the evolution of traffic (Duffe et al 1999). We also cannot find supportive evidence in the public domain of any equivalent legislation or regulation governing the Italian concession. Therefore, we cannot comment as to whether legislative compliance was being maintained by SITMB

Had a service-based analysis been considered prior to the accident we believe the weaknesses in service assurance could have been identified and addressed.

5 Service Analysis Post-Fire

5.1 Services Identified

After the fire there needed to be changes in the operation of the Mont Blanc tunnel given the service failures we have identified (section 4).

Recent papers have confirmed that the basic services required as identified in section 4.1 remain valid (e.g., Bettelini 2022, Lee & Ghazali 2018). Tunnels will always require adequate signage, lighting, ventilation, control and firefighting, and these services, in our opinion, must be assured.

Crucially, the two governments saw the major issues caused by not having a joined-up control structure, so when the Mont Blanc Tunnel re-opened in 2002 a new single maintenance and operations services organisation, The Gruppo Europeo di Interesse Economico del Traforo del Monte Bianco (GEIE-TMB), was formed (Figure 8).

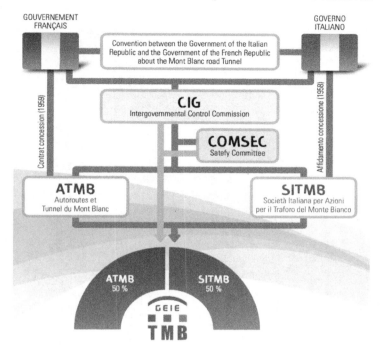

Fig. 8. The Single service structure post 2002 (GEIE-TMB, 2022).

Following our approach to identifying the service hierarchy before the fire, we have determined the breakdown of the GEIE-TMB services is likely akin to Figure 9.

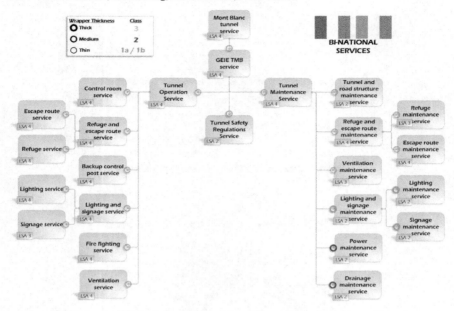

Fig. 9. Service Hierarchy – After the Mont Blanc tunnel reopened post-fire.

5.2 Applying the Service Assurance Guidance

The approach taken for applying the service assurance guidance to the services we believe were extant before the fire (Section 4.2) by considering each service against the Service Objectives that underpin each of the Six Service Assurance Principles remains valid for the services we have identified post-fire.

Equally the concept of LSAs (section 4.2.1) and quality of service assurance evidence (section 4.2.2) remain applicable and were considered in our analyse of the post-fire services. The assessed LSAs and wrappers for the post-fire services are shown on our hierarchy (figure 9). Importantly, our hierarchy of services post-fire shows they are now coordinated under a single organisation.

5.3 Analysis in the context of current operations

Evidence in the public domain (e.g., GEIE-TMB 2022) indicates that the operation of the tunnel has evolved significantly since the fire, and positively concerning safety. A case point being that all users of the Mont Blanc tunnel (and Frejus) are given a dual-purpose leaflet before they are allowed into the tunnel, informing them of "how to use the tunnel safely" and "what to do in an emergency" (figure 10).

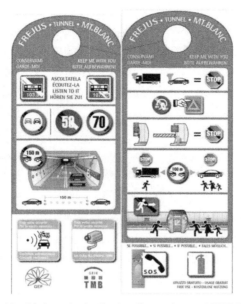

Fig. 10. Double-sided service leaflet for all tunnel users (GEIE-TMB, 2022).

A review of the services as part of the formation of GEIE-TMB and the refurbishment of the tunnel after the fire identified the need to deliver greater assurance through:

- Computerised detection equipment.
- Extra security bays.
- The secure shelters having direct video contact with the control centre, so they can be informed about what is happening in the tunnel more clearly.
- The secure shelters being hermetically sealed and insulated, separated from the tunnel by two fire doors and an airlock system that prevents smoke from entering.
- A separate escape tunnel.
- The safety tunnels have clean air flowing through them via vents, separate to the main tunnel.
- A fire station in the middle of the tunnel.
- Two-headed fire trucks that could travel in either direction within the tunnel without needing to turn around.
- Remote, independently managed sites for safety inspection of all heavy goods vehicles some distance before the entrance on each side of the tunnel. These remote sites would also be used as staging areas, to smooth the flow of commercial vehicles.

These requirements have been addressed as set out in the new Mont Blanc Tunnel Service Charter (GEIE-TMB 2022), with the important safety features as summarised in figure 11.

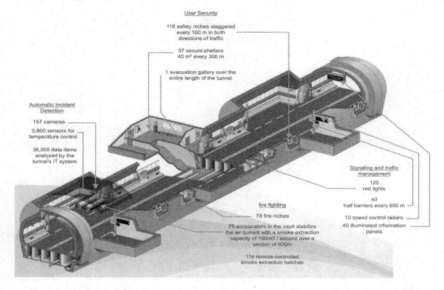

Fig. 11. 2002 revisions to the Mont Blanc Tunnel operation (annotated from GEIE-TMB, 2022)

Supportive evidence regarding the assurance of the improved services is not in the public domain. Therefore, we are not commenting on the assuredness of those services but by applying our knowledge of service assurance (SAWG 2022), we have been able to postulate where we believe the current services (figure 9) could be assured.

Before we detail our analysis, it is important to discuss the European Union (EU) Council Directive that was published in 2004 as this has a direct bearing on the services currently operated in support of the Mont Blanc tunnel.

5.3.1   EU Council Directive

EU Council Directive 2004/54/EC (2004) on *minimum safety requirements for tunnels in the Trans-European Road Network* was devised as a direct result of tunnel fires in Mont Blanc and Tauern in 1999 and St Gothard in 2001. The Directive includes information and requirements on various tunnel services, for example, *Administrative Authority service*, *Inspection Entity service*, *Tunnel Manager service*, *Tunnel Risk assessment service*, etc. It is therefore useful to examine the services described within the Directive to see how they match the new Mont Blanc services which were developed in parallel[7].

The Directive does not state the service hierarchy, but one can infer a hierarchy from it. For example, the inspection entity service will probably be a sub-service of the Administrative Authority as the Directive states:

> "*Member States shall ensure that inspections, evaluations, and tests are carried out by Inspection Entities. The Administrative Authority may perform this function. Any entity performing the inspections, evaluations and tests must have a high level of competence and high-quality procedures and must be functionally independent from the Tunnel Manager.*"

In some cases, the Directive leaves the service structure open, for example:

> "*The Safety Officer may be a member of the tunnel staff or the emergency services, shall be independent in all road tunnel safety issues and shall not be under instructions from an employer in respect of those issues.*"

This means the *Safety Officer service* could be delivered by a member of the Tunnel Manager staff (and thus be a sub-service of the *Tunnel Manager service*) but somewhat confusingly it says the service shall also be independent of the *Tunnel Manager service*.

Our possible hierarchy of services mapped out in figure 9 is analogous to the requirements of the Directive. This is not the only possible service structure that achieves the overall safe tunnel service but would enable a service assurance process to be followed. We believe that it would be useful if future versions of the Directive contained example service hierarchies, to give greater clarity on how services can be assured.

We have found that a service RACI (responsible, accountable, consulted, informed) analysis of the services within the directive is useful and the technique could help provide clarity on the services and how they interact to deliver a safety tunnel service.

Within the Directive there is information on some of the safety requirements for each service, that a service assurance process could use. For example, the *Drainage service* safety requirements are (EU Council Directive 2004/54/EC 2004):

- "*Where the transport of dangerous goods is permitted, the drainage of flammable and toxic liquids shall be provided for through well-designed slot gutters or other measures within the tunnel cross sections.*

---

[7] The Directive entered UK law as Road Tunnel Safety Regulations 2007, with two amendments, Road Tunnel Safety (Amendment) Regulations 2009 and Road Tunnel Safety (Amendment) Regulations 2021, the latter of these severing the link to the EU legislation after the UKs exit from the EU.

- *Additionally, the drainage system shall be designed and maintained to prevent fire and flammable and toxic liquids from spreading inside tubes and between tubes.*
- *If in existing tunnels that requirement cannot be met or can be met only at disproportionate cost, this shall be taken into consideration when deciding whether to allow the transport of dangerous goods on the basis of an analysis of relevant risks."*

We note that the requirements for the *Ventilation service* do not include a similar one, to prevent flammable and toxic vapours from spreading between tubes, a requirement we have met in past tunnel designs.

The requirement that there be only one *Tunnel Manager service* is included in the Directive, but the Directive does allow for other services to be split in cross-border tunnels. The single *Tunnel Manager service* is the same service structure as was adopted by the Mont Blanc tunnel after the fire.

Requirements (supplemented, as necessary) for each service can then be used to review if assurance can be provided for each service, the LSA required, and if service wrappers are needed.

From our analysis of the service failings before the fire, two services were considered substantial contributors to the evolution of the fire and loss of life. The *Control Room Service*, and the *Ventilation Service*. We have taken these services and reviewed publicly available information to see if they can be assured, or not.

5.3.2 Control Room Service

EU Directive requirements for this service (EU Council Directive 2004/54/EC 2004) state:

- *"A control centre shall be provided for all tunnels longer than 3,000m with a traffic volume higher than 2,000 vehicles per lane.*
- *For all tunnels requiring a control centre, including those starting and finishing in different Member States, a single control centre shall have full control at any given time.*
- *Surveillance of several tunnels may be centralised at a single control centre."*

We believe this service to be LSA4. Based on publicly available information, we believe these requirements can be assured.

"Two control and command posts (PCC), which both have the same technical installations, are separately located at the two tunnel entrance aprons, North and South, and are used in conjunction with each other. The one referred to as "active" is under the supervision of two OST (safety and traffic operators) and carries out activities to check the flow of traffic in the tunnel and at the tunnel entrance aprons. The other, referred to as "traffic" (but able to replace

the "active" one at any time), looks after traffic conditions on the access routes. The OST on duty at the "traffic" PCC also has the job of contacting any tunnel users who have reached the safe places." (GEIE-TMB (2022))

Crucially the two Control Rooms at each end of the tunnel are now linked, with an overarching understanding as to how they should coordinate services in the event of an incident, notably a tunnel fire. We would like to hope that such coordination coupled with the design improvements shown in Figure 11, would mean emergency services could reach the site of an incident promptly and not be hampered themselves by inadequate ventilation or immobilised vehicles blocking their passage.

5.3.3 Ventilation service

EU Directive requirements for this service (EU Council Directive 2004/54/EC 2004) state:

- *The design, construction and operation of the ventilation system shall take into account:*
    - *the control of pollutants emitted by road vehicles, under normal and peak traffic flow,*
    - *the control of pollutants emitted by road vehicles where traffic is stopped due to an incident or an accident,*
    - *the control of heat and smoke in the event of a fire.*
- *A mechanical ventilation system shall be installed in all tunnels longer than 1 000 m with a traffic volume higher than 2 000 vehicles per lane.*
- *In tunnels with bi-directional and/or congested unidirectional traffic, longitudinal ventilation shall be allowed only if a risk analysis shows it is acceptable and/or specific measures are taken, such as appropriate traffic management, shorter emergency exit distances, smoke exhausts at intervals.*

We believe this service to be LSA4[8].

"A device for stabilising the longitudinal flow of air, consisting of 76 jet fans placed in the roof, is activated in case of a fire alarm, and allows the longitudinal flow of air to be controlled and facilitates smoke extraction. Smoke extraction is ensured every 100 m by 116 outlets with remote controlled opening, to concentrate the extraction power for sections of 600 m. The smoke extraction capacity at the Mont Blanc Tunnel has been brought up to 156 m^3/sec per 600 m." (GEIE-TMB (2022))

---

[8] It is not generally possible to definitively state the LSA level of a tunnel based on publicly available information, as this requires access to the risk analysis results and information on the specific measures that have been taken to make the risk acceptable.

The vastly improved *ventilation service* that is now in operation covers not just the main tunnel but also the separate evacuation gallery and 37 secure shelters with separate ducting providing the ventilation.

6 Discussion

Had the concept of service assurance been understood prior to the Mont Blanc Tunnel incident and importantly considered by the organisations charged with operating and maintaining the tunnel, we believe the respective governments would have sought change to several of the services in place. Notably in requiring:

- a joined-up approach to communication between the control room services.
- an overarching joint emergency response plan.
- a better understanding of traffic flow compared with the air flow capacity provided by the existing ventilation service.
- a means of communicating with people who may be trapped in the refuge areas.
- greater assurance of the materials being carried through the tunnel and the mechanical state of the vehicles doing that transporting.

However, there is a level of conjecture in here because of the limited pre-fire evidence available to us.

There is greater evidence available since the incident which we have been able to call upon to inform our analysis but even then, this does not consider the services offered pre- or post-fire through the lens of service assurance. Consequently, we have had to make assumptions in our analysis.

The SAWG Guidance (2022) indicates that the criticality of a service should be considered in the form of an assigned LSA with the evidence to assure that service being proportionate to the LSA. Where evidence is deficient the LSA should inform what evidence needs to be constructed in the form of an assurance wrapper. Our analysis shows that the key service failures that contributed to the tunnel fire were of the highest LSA and lacked sufficient assurance evidence either from the service provider or formulated in what could be considered an assurance wrapper by ATMB (or SITMB).

Fig. 12. Post-Fire emergency escape routes

Applying the guidance was able to highlight areas of concern in the service provision that should at least have prompted questions in the governing organisations who were accountable for those services. For example, users of the tunnel were not aware of the risks of remaining in their vehicles as opposed to seeking refuge in the available secure spaces. In the event of an incident the *Tunnel Operation service* should have been able to encourage users to vacate their vehicles and seek refuge. The problem, as Purser (2009) indicates, was that not only were there no such measures in place, but the refuges did not provide a means of escape to the outside and fresh air.

Our analysis indicates the lack of a *co-ordination service* was a significant factor in the way the accident played out and the introduction of an overarching organisation providing such co-ordination after the fire has gone a long way to addressing the service shortcomings we have considered.

One final discussion point which warrants consideration is the trade-off between tunnel safety and operational economics. The new tunnel safety measures (e.g., Figure 12) may be subject to commercial constraints. It is hoped that the independent *inspection service* and much improved regulatory oversight will ensure that GEIE-TMB maintain the managed flow of vehicles through the tunnel and the necessary quarantining and inspection of Heavy Goods Vehicles (HGVs) away from the tunnel entrance.

7 Conclusions and Further Work

Applying the SAWG Guidance (2022) on a real example has proved a valuable exercise in testing its true benefit. However, the limited assurance evidence available for the Mont Blanc Tunnel Fire, notably around the service provision pre-1999, and our need to hypothesise substantially on the service hierarchy, directs to further applications of the guidance to gain confidence in its practicality. Taking this work forward we will now look to identify accident examples where service failure or weakness played a contributory part but there is more assurance evidence available and potentially even an extant service hierarchy that can be critically analysed.

With regard to the application of the SAWG guidance in the early stages of the service lifecycle we must continue to promote its use across the safety-critical systems community. Related to this paper and specific to the safety of services supporting road tunnels, it would seem reasonable that we should offer to engage with the relevant organisations developing the service infrastructure for the forthcoming Stonehenge tunnel and Lower Thames Crossing.

Acknowledgments Many thanks to Mark Sujan for comments on a draft of this paper. Figure 2 is reproduced from flickr.com, https://www.flickr.com/photos/ausdew/38375386084/in/photostream/ All other figures are either reproduced from the respective referenced papers or original diagrams produced by the authors of this paper.

Best endeavours have been made to obtain permissions to use images. In all cases these are taken from public and widely-available sources.

Disclaimer All opinions expressed in this work are those of the authors and not their respective organisations. The analysis presented in this paper has no legal standing whatsoever. The purpose of this paper is not to discredit, contradict or challenge any existing accident analysis; the aim is simply to view the Mont Blanc Tunnel Fire incident through the lens of service assurance. The analysis is the authors' interpretation; they are not speaking on behalf of their employers

References

Bailey, C. (2004), One Stop Shop in Structural Fire Engineering, University of Manchester. https://web.archive.org/web/20181029052004/http://www.mace.manchester.ac.uk/project/research/structures/strucfire/CaseStudy/HistoricFires/InfrastructuralFires/mont.htm (accessed 1st October 2022)

Bettelini, M., (2022) 20 years after the Mont Blanc Tunnel fire – Lessons and legacy, Galleries and Great Underground Works No 142, June 2022.

Catmur, J., Parsons, M. and Sleath, M. (2022) A Step-by-Step Methodology for Applying Service Assurance in Nicholson, M and Parson, M, "Safer Systems: The Next 30 Years", SCSC-170, 2022, https://scsc.uk/r170:1 (accessed 01/10/2022)

Council directive 2004/54/EC on minimum safety requirements for tunnels in the Trans-European Road Network (2004) Official Journal L167, 30.4.2004, p. 39.

Duffe, P., Marec, M. and Cialdini, P. (1999) Joint report of the Italian and French administrative commissions of enquiry into the disaster that occurred on 24 March 1999 in the Mont Blanc Tunnel. République Française, Ministère de l'intérieur et Ministère de l'Equipment, du Transport et du Logement, Repubblica Italiana, Ministero dei Lavori Pubblici, 6 July 1999.

Directive 2004/54/EC of the European Parliament and of the Council of 29 April 2004 on minimum safety requirements for tunnels in the Trans-European Road Network, https://eur-lex.europa.eu/legal-content/EN/TXT/?uri=celex:32004L0054 (accessed 01/11/2022)

Durston, N., Parsons, M., Scott, A. and Simpson, A. (2019) The Principles of Service Assurance in Kelly T and Parsons M, "Engineering Safe Autonomy", SCSC-150, 2019, https://scsc.uk/rp150.6:1 (accessed 1st October 2022)

Elliott, S. and King, K. (2019) Service-based Safety Assurance: A provider approach in a challenging environment in Kelly T and Parsons M, "Engineering Safe Autonomy", SCSC-150, 2019, https://scsc.uk/rp150.6:1 (accessed 1st October 2022)

GEIE-TMB (2022) The Mont Blanc Tunnel Service Charter https://www.tunnelmb.net/public/files/301/carta-dei-servizi-ing-2022.pdf (accessed 06/11/2022)

GEIE-TMB (2022) One tunnel, two countries, fifty years of history, https://www.tunnelmb.net/en-US/history (accessed 1st October 2022)

Harris C, Parsons M and Simpson A (2018) Service-Based Safety Assurance in Kelly T and Parsons M, "Evolution of System Safety", SCSC-140, 2018, https://scsc.uk/r140/8:1 (accessed 1st October 2022)

IfM and IBM. (2008). Succeeding through service innovation: A service perspective for education, research, business and government. Cambridge, United Kingdom: University of Cambridge Institute for Manufacturing. ISBN: 978-1-902546-65-0

King, K., Parsons, M. and Sujan, M. (2020) A Service Perspective on Accidents in Nicholson, M and Parson, M, "Assuring Safe Autonomy", SCSC-154, 2020, https://scsc.uk/r154:1 (accessed 1st October 2022)

Lee, Y. S. & Ghazali, F. (2018). Major Functional Risks for Operation and Maintenance of Tunnelling Projects. ESTEEM academic journal. 14. 24-37.

Purser, D. (2009). Application of human behavior and toxic hazard analysis to the validation of CFD modelling for the Mont Blanc Tunnel fire incident. In: Proceedings of the Fire Protection and Life Safety in Buildings and Transportation Systems Workshop. 23-57.

Road Tunnel Safety Regulations 2007 (SI 2007/1520), Available at: https://www.legislation.gov.uk/uksi/2007/1520/made (accessed 06/11/2022)

Road Tunnel Safety (Amendment) Regulations 2009, Available at: https://www.legislation.gov.uk/uksi/2009/64/regulation/made (accessed 06/11/2022)

Road Tunnel Safety (Amendment) Regulations 2021, Available at: https://www.legislation.gov.uk/uksi/2021/552/regulation/made (accessed 06/11/2022)

SAWG (2020), Service Assurance Guidance v1.0, SCSC-156, published at SSS'20, February 2020, https://scsc.uk/r156:1 (accessed 01/10/2022)

SAWG (2021), Service Assurance Guidance v2.0, SCSC-156A, published at SSS'21, February 2021, https://scsc.uk/r156A:1 (accessed 01/10/2022)

SAWG (2022), Service Assurance Guidance v3.0, SCSC-156B, published at SSS'22, February 2022, https://scsc.uk/r156B:1 (accessed 01/10/2022)

Tarada, F. (2021), Road Tunnel Safety After the Mont Blanc Fire, Presentation to the Safety and Reliability Society, 30th March 2021, https://mosen.global/wp-content/uploads/2021/04/Road-Tunnel-Safety.pdf (accessed 01/10/2022

Legal Liability for Safety Critical Systems

Dai Davis

Percy Crow Davis & Co

Abstract   *Liability for defective systems arises in both criminal and civil law. There are no significant differences or exceptions to liability attaching to defective software within systems. The civil liability sets a higher bar than the criminal system: to put it another way if a manufacturer produces a product which is safe enough so as not to attract civil liability if it goes wrong, that manufacturer will certainly have no criminal liability if the product is nevertheless defective. Dai Davis, who is both a Chartered Engineer as well as a Solicitor will explain precisely what liability attaches to defective software and more importantly what a manufacturer or software author must do to ensure that a product (or the code) is "safe" from a legal perspective. He will explain in detail the civil legal wrong of Product Liability as well as looking briefly at the criminal law offences relevant to unsafe products. Finally, he will examine the issue of personal liability as opposed to corporate liability.*

© Dai Davis 2023.
Published by the Safety-Critical Systems Club. All Rights Reserved.

ns and Safety Engineering. It
Challenges in Safety Mitigation and Assurance of Multi-agent Complex Systems

Alastair Faulkner[1] and Mark Nicholson[2]

[1]Abbeymeade Limited, [2]University of York

Abstract *Systems composed of many interacting components are complex. This complexity varies with scale, the number and the type of active members. Complex system behaviour is intrinsically challenging to model due to the dependencies, competitions, relationships, or other interactions between their parts or between a given system and its environment. Historically, participants were users; in future systems, software agents will likely undertake these roles. Other properties might arise from agent relationships, such as nonlinearity, emergence, spontaneous order, adaptation, and feedback loops. This emergence of technological agents begs the question of how should we mitigate the potential safety effects of agent failures? How can we assure ourselves that an agent that takes an active role in a system is adequately safe? What if there are multiple interacting agents? This paper considers the use of 'Engineered Agents' and how they might change the activities undertaken by Systems and Safety Engineering. It addresses the nature of single-point and common-cause failures. How might these failure modes be addressed and assured for agent-based systems? Finally, the paper addresses architectural matters and indicates how those issues contribute to safety assurance and its associated approvals in technological agent-based systems.*

Keywords: Agent-based systems, Engineered Agents, common cause failures, Data, Metadata, Metamodels.

1 Introduction

Computational platforms have advanced so that a single physical platform is no longer a primary consideration in systems engineering and systems safety. Such

---

[1] email: alastair.faulkner@abbeymeade.co.uk

[2] email: mark.nicholson@york.ac.uk

© Alastair Faulkner and Mark Nicholson 2023.
Published by the Safety-Critical Systems Club. All Rights Reserved.

advances are evident in the Internet of Things (IoT), where general-purpose operating systems form the basis of communications-enabled devices. A similar evolution is apparent in mobile phones and their applications. Software agents have evolved from their most notable roots in algorithmic (financial) trading, also known as an automated trading system. These systems detect changes in the trading in the financial markets; when they get it wrong, the effects can be spectacular.

On 6 May 2010, a United States trillion-dollar flash crash (a type of stock market crash) started at 2:32 pm EDT and lasted approximately 36 minutes (Wiki 2010 flash crash). The most likely cause was the detection of 'rates of change' in the market, triggering the automated trading systems to sell. This was not the result of actions by one agent. The combination of many heterogeneous agents implemented on many platforms with complex interdependencies and relationships, acting independently and asynchronously, led to positive feedback creating a crash.

The potential for "crashes" with significant impact arises as the use of agents, and multiple agent systems moves into the engineered system domain where a failure may have an impact on system safety. As a result, in this paper, we develop the concept of an 'Engineered Agent' (EA) as a systems element with independent and group service provision capabilities that is valid in a context. The simplest forms of EA are stand-alone and static, with a limited set of known (specified) behaviours, capacities, and capabilities. As the system becomes more complex, the EA may employ strategies to refine its responses based on past performance. We then address the safety engineering and assurance aspects of multi-agent complex systems.

2 Defining an Engineered Agent

An **agent** is an **entity** situated in some **environment** capable of autonomous **action** in this environment to meet its objectives (Siegfried 2014)[1]. Siegfried supports his definition with the following:
- Agents are identifiable, discrete (and usually heterogeneous) individuals (North and Macal 2007, p. 214).
- Agents are space-aware, i.e., situated in some environment (Macal and North 1983, Borshchev and Filippov 2004, Siebers and Aickelin 2007).
- Agents are capable of autonomous action and independent decisions. In this sense, agents act actively rather than purely passive objects (Wooldridge 1999, p. 28ff., North and Macal 2007, p. 214, Klugl 2006).

---

[1] Siegfried states that there is no accepted definition of an agent (Siegfried 2014). Siegfried is concerned with modelling and simulation and not the broader context of system engineering. His book does, however, provide a useful starting point.

- To act within the environment and pursue their goals, agents can perceive and work within it (Klugl et al 2006, Wooldridge 1999, p. 32).

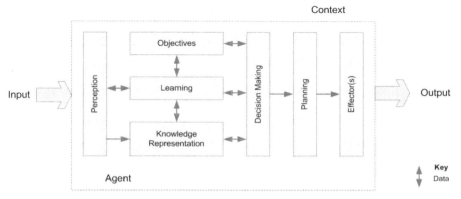

Fig. 1. Agent in Context (Hajduk et al 2019)

Figure 1 redraws a generalised representation of an agent in a context (Environment (Siegfried 2014) based on Hajduk (Hajduk et al 2019).

Each agent is provided with a stimulus from its environment. It reacts to that stimulus by following a set of internal goals. To do this, it must interpret the stimulus by creating an internal representation of the information supplied by the stimulus. It then compares the information with its objectives and decides whether it needs to change how it interacts with the world to maximise conformance with its objectives. In parallel, it judges the impact of its previous actions and environmental changes. It uses this information to learn whether to change its objectives, the set of available decisions, or the value of each available decision. Having decided, it addresses the action phase of its operation. It plans the action it has decided to take and implements this action. The output is a control action that maintains or alters the way the agent interacts with its context (environment).

Hajduk provides an initial list of agents:
- **re-active agent** - always perceives its environment and responds in a timely fashion to changes that occur in it to satisfy its design objectives.
- **model-based reflex (intentional) agent** - considers its options to achieve its goal.
- **hybrid agent** – combines both perspectives as mentioned above.
- **agent with behaviour-based architecture** - employs a collection of concurrently executing behaviours and decoding models of behaviour. In contrast, each behaviour model represents a goal-oriented block responsible for achieving a specific task.

This list is extended by additional requirements identified later in this paper.

2.1 Additional Requirements for the Agent Definition

The environment is better stated as a context. In this way, many agents may be active in an environment. The degree of their context's overlay (intersection) becomes the subject of the architecture and design at the next higher level of system abstraction (see Figure 2).

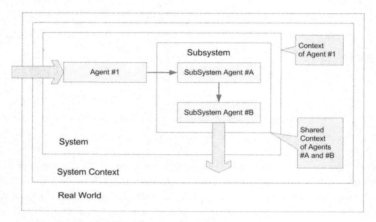

Fig. 2. Agent and System Context (Environment)

The development and evolution of system architecture is an emergent property because of the changes each agent may make in its objectives, set of decisions and the value of each decision. However, engineering principles and techniques are required to guide and constrain emergent properties. The interconnectedness of contexts allows strategies for managing joining (merging of) and divesting (separation of) agents in a system.

Here we define **Context** as any information used to characterise the situation of an entity (Dey and Abowd 2000). We can qualify this definition, so **System Context** means the highest-level view of a selected System of Interest (SoI) and its relationships in its environment (Flood and Carson 1993). The system context should relate an agent's viewpoint to the relevant elements of the context (SE-BoK2016Context).

Therefore, EA may be arranged as a hierarchy (Figure 3 (Hajduk et al 2019)) and, consequently, be subject to supervision and monitoring by agents designated to undertake this role. These hierarchies would allow the application of standard safety engineering patterns of monitoring and protection architectures. However, the elements of the patterns will need to be suitably amended to address the characteristics of a multiple EA environment.

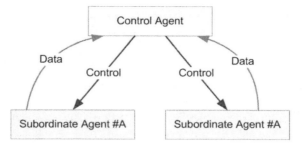

Fig. 3. Multi-agent Architecture

An EA will be a decision-maker. A common, perhaps misguided, reuse of the term autonomy implies that agents have the capacity for self-government, a form of independence with limited or no oversight or constraint. Within a hierarchy, an EA will be heteronomous[2] if its "will" is under the control of another agent. Thus, command and control concepts are appropriate when addressing how a multi-agent system will respond to its environment.

Not all EAs are equally important for the provision of a service or the provision of safety mitigation in a multi-agent EA system. Design patterns such as Observe, Orient, Decide, Act (OODA) (Shahbazian et al 2001), Perception, Comprehension and Projection (PCP) (Rowan 2021, Endsley 1995, Sandom 2017), and Sense-Understand-Decide-Act (SUDA) have been developed to represent the activities to be undertaken by an agent. The degree to which an EA implements the SUDA, OODA or PCP triplet determines the primacy of the EA in the overall system architecture.

The creation, design, development, and release of EA-based systems is a starting point rather than a conclusion. These systems require upgrades and maintenance. Functions, capabilities and constraints will all change. As a result, the question of how Configuration Management and Change Control will be implemented is pertinent.

A pragmatic set of Configuration Items will not only include the set of EAs and its operational context but will also include defined interfaces (managed through Interface Agreements (IA)) and the data and controls exchanged over the IAs. As a result, in complex EA-based systems, it will be essential to apply configuration management to metadata and metamodels, as these characterise the IAs and allow assurance activities to take place over them.

---

[2] Autonomy is the capacity for self-government. Agents are **heteronomous** if their will is under the control of another. Oxford Reference (Dictionary of Philosophy)

2.2 Multi-agent Systems

Given the widespread use of IoT and autonomy (and heteronomy), agent-based systems are in an ever-expanding installed base. New agents from multiple vendors, different versions, capabilities and capabilities will produce a rich, diverse heterogeneous context (Jezic et al 2019).

We do not suggest the wholesale replacement of existing agents. We require a suitable and sufficient safety analysis to identify those system areas with safety responsibility. If we were to implement an industrial process plant and its associated equipment, we would use a hierarchy of components to implement the safety systems. For example, we would use IEC 61511 accredited Programmable Logic Controllers (PLC) for the plant logic and safety protection systems. This high-integrity system would interface with the control equipment to provide the user workstation (for oversight). Interfaces with non-safety systems would provide data, schedules and status information. This hierarchy is good practice and provides the bases for a security model and its enforcement of access control for authentication and their associated authorisation.

The safety-accredited PLC would have addressed issues associated with Single-Point-of-Failure (SPoF) and Common Cause Failures (CCF). These are managed through architecture and accredited toolsets (workbenches). EAs require similar accredited development practices and toolsets.

An Engineered Agent **acts** in a defined **context** within Systems of Systems (SoS). An EA is capable of autonomous (or heteronomous) action within its context to meet its **objectives**. An EA has:

- Defined functionality, capacities, capabilities and constraints.
- Defined boundary, typically with interfaces based on **Interface Agreements** (IA). An IA is a set of one or more constraints which system components must uphold to meet one or more requirement(s) (Faulkner and Nicholson 2020).
- Some degree of **learnt behaviour** ranging from parameter refinement to Machine Learning (ML).
- **Context-aware** to perceive relevant information and receive services or commands. The relevancy of these services and commands depends on the EA's objectives.
- **Decision-making** heavily relies on context, its maturity, perception of the current state of the system and its environment, alternatives available, and the training, experience and competence of the decision-maker or group (Kahneman and Tversky 2000) for its correctness.
- Capable of stand-alone, collective and hierarchical configurations.

Implementing a safety-related multiple EA system requires consideration of diversity, including computational platforms, communications, connectivity, architecture and data diversity. Systematic error may exist in the data representation, model, metadata and metamodels. Care is required to identify unconscious bias and implicit assumptions of use.

A system employing multiple identical agents is homogeneous. They are Consistent, Conformant, and Compliant. They depend on shared resources and have a common purpose and mutual interdependence. However, they may have different contexts. They will have other inputs and therefore learn independently.

The System Definition remains a crucial document defining the overall system boundary, subsystems, internal and external boundaries, Interfaces and IAs.

3 Complex Systems

A Complex Adaptive System (CAS) is a system where the individual elements act independently but jointly behave according to shared constraints and goals (Weaver 1948; Jackson, Hitchins and Eisner 2010; Flood and Carson 1993; Lawson 2010) (BKCASE).

Complex systems contain many elements that lead to dependencies, competitions, relationships, or other interactions between elements or a given system and its environment. These systems become CAS when new elements and changes in the elements' behaviour cause other elements to change their behaviour – the overall system is adaptive with new sets of emergent properties.

Figure 4 provides a way to visualise the spectrum of Complexity versus Autonomy in a multiple EA system. Figure 4 is intentionally blank; there is value in the uncertainty, requiring the reader to ask what the relationship is. Contemplating the figure requires the reader to try to place examples of AI on the figure.

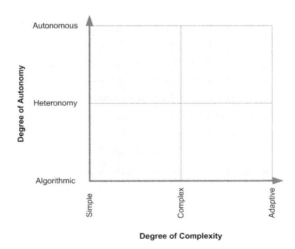

Fig. 4. Relationship between Complexity and Autonomy (intentionally blank)

In large distributed systems, all components are unlikely to be in a 'normal' operational mode. Therefore, a pragmatic consideration of the impact of degraded and emergency modes and accidents on the safety properties and safety assurance is also required.

Further consideration of degraded modes should address the maintenance of the elements of the multiple EA systems. For example, how will 'permit-to-work' (PtW)[3] be implemented? What form will lockout devices[4] take, and how will this be instantiated? More importantly, how will their removal be authorised?

Authorisation requires a security model for accreditation and subsequent access and authority while undertaking the maintenance action. Authorisation implies 'identity'. Will each EA need an identity? Can an EA be trusted? What confidence can we have in the EA as a decision-maker? What evidence is required to demonstrate the correct operation of the EA and support the investigation?

Finally, how will the EA implement perception of its context in a way that allows the ground truth of the context to be used to make an action or learning decision? Direct connection to sensors (as strong data) presents few difficulties. In contrast, weak data is indirect, derived from sources such as links to third parties or reliance on one or more data chains that contain many transformations. Therefore, the weight that each data source is given in an agent's perception algorithm is key to the actions an agent takes.

4 Systems and Safety Engineering

What could possibly go wrong? Errors, faults and failures will occur in all system elements, including EAs. The EA's emergent behaviour is not confined to software but is influenced by the data it consumes. Through perception and comprehension, the agent will learn and adapt its behaviour accordingly within specified bounds. At some unspecified time, the EA will use the learnt behaviour. It may, of course, learn inappropriately. How might these failure modes be addressed and assured for multiple EA-based systems?

---

[3] "Typically, a PtW is a management procedure where only persons with specific authority will sign a permit on which ostensibly the life of a worker might depend. To this end, responsibility for the PtW rests with the person in charge of the operation for which the permit is required." "The CLEARANCE of a permit after the job is done is as important as the RAISING of it.)" IET: Permit to Work Systems (Health and Safety Briefing No. 33)

[4] Lockout device – used in conjunction with PtW, in physical systems used to let employees know that essential precautions have been taken and, where necessary, physical safeguards are in place - e.g. circuits open and contactors or switches securely locked open. Equivalent practices are required to support change in abstract entities – such as the definition of a segment of airspace. (Faulkner and Nicholson 2020)

The arrangements of the software, hardware and communications elements remain core architectural considerations as they address SPoF and CCF. However, as systems become more complex, an abstract architecture is essential. One approach is to use Model-Based Systems Engineering (MBSE) to formalise the abstract conceptual constructs' and, therefore, to provide management of the consistency of definition and use.

However, a single unified and integrated MBSE model may require iteration to provide the basis for safety assessment and analysis. Practical systems are likely to be based on several MBSE models. Large-scale systems typically contain legacy equipment which predates the MBSE models. The challenge will be to conduct a suitable and sufficient assessment and analysis, especially in adaptive systems with many emergent properties.

Tool support is vital to enforce consistency and constraints. Competently used, the tool will embody the documented management strategy, its user manual, and a plan for the MBSE models and their integration in a complex domain based on Systems of Systems (SoS). This paper assumes that the computational platforms are resilient, protected from cyber security issues and powered by uninterruptable power supplies.

Architectural safety requires segregating high-integrity functionality into identifiable bounded entities to protect from errors in lower-integrity elements. Therefore, safety assurance addresses the safety justifications for this high-integrity functionality, including identifying and managing the Safety-Related Application Conditions (SRAC). We need to be careful to avoid the trap of providing safety assurance for equipment collections rather than systemic abstraction.

Where mixed integrity functionality shares a platform, failures of low-integrity functions must be independent of the high-integrity safety elements. Maintenance access via 'back channels' provides a way to undo the intended segregation. Agent-based architectural patterns should address monitoring, redundancy and diversity. Communications to and between agents require strategies to handle the loss, and subsequent restoration, of communications. What learnt behaviours have been created during the communications outage, and are they still valid on restoration? Will these behaviours need to be removed and the earlier system state restored? What emergent properties have arisen and why? These are considerations that are familiar to competent safety practitioners.

Further, the response of the EAs in the system to failures needs to be addressed. What strategies could be used, and by which agent? For example, one standard safety-critical architectural pattern is the control-monitor architecture. Here the monitor provides alarms. In multiple EA systems, alarm avalanches can occur. Questions such as how alarm avalanches are managed, will alarms be prioritised, and what entity will receive these alarms must be addressed. If the entity receiving an alarm is an agent, how will this agent manage its response? Will it set one or more system elements to degraded mode and plan around them? How will the normal operation be restored?

Another question is how will Configuration Management (CM) and Change Control (CC) with respect to existing safety requirements and new emerging safety requirements be implemented in agent-based systems to ensure the safety properties of the multiple agent system? How will this be addressed in the confidence case associated with the EA-based systems safety case? Where EA behaviour is static, known and predictable, CM will use existing approaches. In complex systems, using multiple EAs in a hierarchy, or operating as a loosely coupled collection, interdependent and interrelated, then properties might arise from these relationships, such as nonlinearity, emergence, spontaneous order, adaptation, and feedback loops. This implies that CM strategies are required to implement dynamic Configuration Audits. When challenged, the critical question will be 'at the time of the incident, what was the composition of the system'?

In addition, EAs will have failure modes associated with learnt behaviour, decisions, and relationships. Existing techniques will require review to extend their scope. For example, the Fault Tree Analysis (FTA) pattern Primary, Secondary, and Command should be expanded to include 'Decision', addressing decision-making failures. A CAS co-evolves with its context (Dodder and Dare 2000). CAS have failure modes related to nonlinearity, emergence, spontaneous order[5], adaptation, and feedback loops. Currently, EA component-level analysis will not expose these failures.

And finally, how will PtW be implemented where agent-based systems connect to real-world devices? How will the lock-out devices be applied? Which agents are authorised to remove the lock-out device and undertake the managed return-to-service? How will unique identities and authorities be enforced? How will imposters be detected? Will the agent-based system detect cyber-attack and intrusions? Here there is an overlap with cyber-security.

5 Example Application Areas

New transport vehicles illustrate a shift away from established technologies based on hydrocarbon use. In the automotive domain, these vehicles use hybrid and all-electric technologies.

5.1 All-Electric Transport Platform

Consider a generic All-Electric Transport Platform (AETP) that includes a power source (typically a battery), sensing technologies, propulsion, braking, direction

---

[5] Spontaneous order – using "system thinking" to describe dynamic, self-organising networks of structured interactions.

control (steering and braking) and one or more EAs. For this example, each EA executes PCP, subsequent decision-making and action to interact with its context, including other EAs.

A context is required if the AETP is to operate safely and successfully. Navigation is critical. Therefore, a definition of the operational space, obstacles and, where appropriate, planning (navigation) rules. Additional detail is required for start and stop locations as these typically involve the most significant degree of interaction with other AETPs.

5.2 Air Taxi

Unmanned Aerial Vehicles (UAV AETP) operate either remotely piloted or autonomously. This functionality will include navigation avoiding obstacles, take-off, flying and landing of planned journeys. These vehicles will employ EA; these EAs will communicate with intelligent infrastructure EAs and be monitored by EAs.

Besides the existing built environment (buildings et al.) and terrain, a single UAV requires a minimum infrastructure set. We can construct a safety argument for a single UAV based on past aviation experience using the airworthiness of the UAV, its failure modes, its endurance, operational constraints and SRACs.

Multiple UAVs are a more complex safety problem. Interaction between the UAV EA, other UAVs, and the infrastructure will expose dependencies and interrelations that are both direct and indirect. In some cases, strong data (from sensors) may be replaced with weak data derived, transformed, abridged and consolidated through third parties. One approach is to use definitions shared by all UAVs; for example, common flight corridors are defined to reduce navigation conflicts. These corridors intersect at nodes. This definition includes a set of rules for navigation manoeuvres to reduce the risk of collision.

The UAV system quickly becomes a Multi-Agent Complex System.

5.3 Social Agents

A social agent interacts with other agents through communication. Where there is mutual benefit, an EA may coordinate and cooperate with other EAs. For example, UAVs may combine into groups to optimise throughput at a navigational node. An emergent property would be the shared sensor data enhancing perception. This improved perception may enable a degraded UAV to 'tag along' with a group when non-essential sensors have failed.

Social attractions may allow highly capable EAs to form hierarchies. The group example would allow new UAVs (with the latest EAs) to control and, therefore, enhance the capabilities of an older, by implication, less functional UAV. This possible assistance introduces new failure modes for the controlling UAVs EAs to address. What happens to the assisted UAV?

5.4 Complex Adaptive Systems

A UAV CAS would be a mass transit system. Consider multiple manufacturers providing multiple UAVs (including an installed base of different versions) and numerous operating companies giving competitive rates for the customised journey. Each UAV and its EAs act independently and jointly; they comply with shared constraints and goals.

We need only look at current road vehicles and proposed autonomous cars to observe the infringement of the inviolate 'rules of the road' to create aggressive behaviour at junctions. These infringements create uncertainty in other road users, rewarding aggressive driving. How might a UAV implement a bold flying style?

Collisions between UAVs are likely to have higher consequences. UAVs do not need to collide to create an incident. A UAV may take a shortcut; the propulsion wake is sufficient to induce turbulence and flight instability. Thus, even minor flight infringements enhance existing flight hazards and, therefore, more significant risk.

High-traffic, high-density UAV system compliance to shared constraints and goals is vital. How could this compliance be enforced? One approach requires the UAV to monitor and record its performance and the performance of other UAVs. Groups of local UAVs provide infringement detection to the errant UAV. Enforcement requires the withdrawal of the flight authority. Flight authority implies a centralised flight controller, which itself might be an EA.

Very quickly, the mass transit UAV CAS exhibits nonlinearity, emergence, spontaneous order, adaptation, and feedback loops.

6 Discussion

Multi-agent complex systems are not new. The replacement of the human as the agent by software agents is increasingly common. All agents are not equal. Agents with safety responsibility should be designed, developed, approved, installed and maintained in the spirit of all other safety systems.

This paper has described examples where agents are used and expose safety issues that need to be addressed. There is a clear case for 'Engineered Agents'. Further, the assurance, and confidence around this assurance, also need to be addressed. Using EAs will lead to changes to the activities undertaken by Systems and Safety Engineering.

Domain-specific safety approvals practices require review and revision to adapt the agent to the use of EAs and EA-based technologies. Safety practitioners should structure their approach to set out policy-strategy-plan issues. This is critical where EAs implement permit-to-work, manage lock-out devices and undertake the managed return-to-service.

Using generic platforms and communications increases the likelihood of SPoF and CCF. Generic platforms also present challenges of diversity. Architectural patterns are required to support the implementation of EAs to ensure fail-safe EA systems designs. These architectural patterns do not currently exist. So what architectural patterns could agent-based systems employ? One starting point is to consider independence, resilience, diversity and the management of SPoF and CCF. EA functionality should include diagnostic and health check functions.

Further work is required to develop patterns of agent-based systems:

- Control (Single channel, Dual Channel and voting systems)
- Monitor, logging and alarm oversight
- Collective (groups, swarms and collectives)

Many of these patterns will include communications-dependent technologies, requiring assessment to identify constraints. Further, for EAs to interact appropriately at run time, IAs between EAs will need to incorporate identity and build status handshakes so that the ability to interface can be established so that actions, or joint actions, can be created. Sometimes, a poor build state of an agent may make it appropriate to isolate it until it has undergone a round of maintenance.

By omission, we have implied that each EA is a resident on an identifiable hardware platform. Given the ubiquity of hardware and software platforms, EAs can execute across the IT estate.

It rapidly becomes clear in this context that scale, scope and complexity are significant challenges for CM. A strategy requiring EAs to return their identity and build status provides a minimum basis for configuration audit. Establishing the build state during a safety incident is an essential first step for incident investigation.

Physical incident investigation relies on establishing witness marks in failed components. These may be fatigue fractures or impressions of parts as fragments on surfaces. Identifying the state of the SoS at the time of the incident is essential but may need more evidence than is currently connected. The incident sequence may include many EA decisions by many EAs. Digital evidence will be required to establish the basis of decision-making by an EA and its potential contribution to the incident. Further, learnt behaviours, weighting and bias may also be relevant.

Incident investigation will require a simulator to show the evolution of the interactions between agents. In the ideal, that simulator will be provided by an independent third party. The simulator may be in a different legal jurisdiction, rendering attempts to retrieve simulator results impeded by legal argument.

Even minor incidents may require large volumes of data to analyse. Consider an incident in the UAV system where a group of UAVs infringes on an airspace constraint. Where is the logged data of this infringement and its potential causes held? How are the timestamps coordinated across the EA-based system? How will emergence, spontaneous order, adaptation, and feedback loops be recorded? What permits-to-work are relevant to the incident sequence that might have led to maintenance-induced failures? How were lock-out devices managed? Which EA undertook the managed return-to-service?

The use of multiple, possibly competing, EAs are illustrated by adapting the Byzantine Generals Problem. This describes decentralized EAs' difficulty in arriving at a consensus without relying on a trusted central party. No EA can verify the identity of other EAs. The question therefore arises how can EAs collectively agree?

Even our simple explanation of the UAV, multiple EAs and its potential use of groups illustrate how complexity grows - existing techniques manage complexity through modularity. One approach is to combine modules into collections, often with a hierarchy. These collections have defined boundaries that provide barriers to escalation, reducing the potential to propagate errors, faults and failures.

In the UAV example, we have yet to explore identity as a mechanism to identify an EA. Identity provides a means to authenticate an EA and assign access and authorities. The UAV concept of free movement via direct flights is diminished by the requirements of multiple UAVs and safety constraints.

We propose extending the list of agents proposed by Hajduk (Hajduk et al 2019) to include at least:

- **Supervisor Agent**: to act as 'group manager' and for other generic All-Electric Platform (AEP) as the supervisor's subordinate.
- **Command Agent**: to act as driver or pilot executing the planned journey
- **Federated Sensor Data Agent**: the management of local data and requests for data from others in the group.

Should this list include EAs specifically for Perception, Comprehension and Projection? Perhaps a specific class of EAs for compliance?

7 Conclusions

The generic AEP example illustrates multi-agent systems' almost unbounded potential for complexity. The case for constraining and managing this complexity using Engineered Agents is founded on existing safety engineering and operational practices. Safety analysis should not be applied differently simply because EAs are implemented on different technologies. However, variations of approach, such as the use of interface agreements and a more dynamic configuration management process, may well result from implementing such systems. One of the most potent questions remains, "how could EA-based systems possibly go wrong"?

In this paper, we argue that existing systems labelled as autonomous are more likely to be a collection of EAs. Separating these EAs allows for analysis and may lead to improved, more robust architectures and resilient systems.

Agents are not new. Human agents operate many categories of systems. Many of these human agents are currently being replaced by software agents. Human agents become highly competent and combine perception with anticipation. The human agent has the potential to adapt dynamically to unusual circumstances and operational conditions. Is the use of agents a better way to architect for the increased decision-making and action response than currently proposed autonomous architectures?

Overall, this paper illustrates that research is required to address the architecture of EAs to identify SPoF and CCF and to create strategies to address them at run time. Safety assurance needs to evolve to address the urgent issues associated with the operational properties to expose relationships, such as nonlinearity, emergence, spontaneous order, adaptation, and feedback loops. Using data safety approaches, such as interfaces and, dynamic IAs, may provide good strategies to address these issues.

References

Andrei Borshchev and Alexei Filippov. (2004) From System Dynamics and Discrete Event to Practical Agent Based Modeling: Reasons, Techniques, Tools. In The 22nd International Conference of the Systems Dynamics Society, Oxford, England

BKCASE Governance and Editorial Board. Guide to the Systems Engineering Body of Knowledge (SEBoK). 2017. URL: http://sebokwiki.org/wiki/Complex_Adaptive_System_(CAS)_(glossary)

Anind K. Dey, and Gregory D. Abowd, (2000) Towards a better understanding of context and context-awareness. Proceedings of the Workshop on the What, Who, Where, When and How of Context-Awareness, affiliated with the CHI 2000 Conf. on Human Factors in Computer Systems

Rebecca Dodder and Robert Dare, (2000) Complex Adaptive Systems and Complexity Theory: Inter-related Knowledge Domains, ESD.83: Research Seminar in Engineering Systems Massachusetts Institute of Technology

Emerson, Matthew, and Janos Sztipanovits. (2006) "Techniques for metamodel composition." OOPSLA–6th Workshop on Domain Specific Modeling.

Mica R. Endsley. (1995) Toward a Theory of Situation Awareness in Dynamic Systems. Volume 37(1). Human Factors - The Journal of the Human Factors and Ergonomics Society, pages 32–64

Alastair Faulkner and Mark Nicholson (2020). Data-Centric Safety: Challenges, Approaches, and Incident Investigation. Elsevier.

Robert Flood and Ewart Carson (1993), Dealing with complexity: An Introduction to the Theory and Application of Systems Science, 2nd ed. Plenum Press, New York, NY, USA, 1993, 978-0306442995

Gartner, Richard. (2016) Metadata. Springer.

Mikuláš Hajduk, Marek Sukop, Matthias Haun, (2019) Cognitive Multi-agent Systems Structures, Strategies and Applications to Mobile Robotics and Robosoccer, Springer, ISBN 978-3-319-93685-7

Jackson, Hitchins and Eisner, Jackson, Scott and Hitchins, Derek and Eisner, Howard, (2010) What is the systems approach? INCOSE Insight 13:(1)

Gordan Jezic, Yun-Heh Jessica Chen-Burger, Robert J. Howlett, Lakhmi C. Jain Ljubo Vlacic Roman Šperka (Editors) (2019) Agents and Multi-Agent Systems: Technologies and Applications 2018, Proceedings of the 12th International Conference on Agents and Multi-Agent Systems: Technologies and Applications (KES-AMSTA-18), Springer

Kahneman, D. and Tversky, A. (2000), Choices, Values, and Frames, Cambridge University Press

Franziska Klügl. (2006) Multiagentensimulation. Informatik-Spektrum, 29(6):412–415, December 2006.

Michael E. Kuhl, Natalie M. Steiger, F. Brad Armstrong, and Jeffrey A. Joines, editors. (2005) Proceedings of the 37th Winter Simulation Conference, Orlando, FL, USA, December 4-7, 2005. ACM, December 2005.

Lawson, Harold, (2010) A Journey through the Systems Landscape, College Publications, Kings College, 2010, ISBN 978-1-84890-010-3

Charles M. Macal and Michael J. North. (1983) Tutorial on agent-based modeling and simulation. In Kuhl2005, pages 2–15.

Michael J. North and Charles M. Macal. (2007) Managing business complexity: Discovering Strategic Solutions with Agent-Based Modeling and Simulation. Oxford University Press, New York.

Rowan, Wendy, et al. (2021) "Comprehension, Perception, and Projection: The Role of Situation Awareness in User Decision Autonomy When Providing eConsent." Journal of Organizational and End User Computing (JOEUC) 33.6 : 1-31.

Shahbazian, Elisa, Dale E. Blodgett, and Paul Labbé. (2001) "The extended OODA model for data fusion systems." Fusion01.

Sandom, Carl (2017) Automation Autonomy and Awareness, Safety Critical Systems Club - Seminar: New Thinking in Human Factors for Safety https://scsc.uk/file/480/03---Carl-Sandom---Automation-Autonomy-and-Awareness.pdf

SEBoK2016Context: BKCASE Governance and Editorial Board,Guide to the Systems Engineering Body of Knowledge (SEBoK) https://www.sebokwiki.org/wiki/System_Context_(glossary)

Shahbazian, Elisa, Dale E. Blodgett, and Paul Labbé. (2001) "The extended OODA model for data fusion systems." Fusion01.

Peer-Olaf Siebers and Uwe Aickelin. (2007) Introduction to Multi-Agent Simulation. Preprint for encyclopedia of decision making and decision support technologies, University of Nottingham, 2007. http://eprints.nottingham.ac.uk/645/.

Robert Siegfried, (2014) Modeling and Simulation of Complex Systems: A Framework for Efficient Agent-Based Modeling and Simulation, Springer Vieweg, ISBN 978-3-658-07528-6

Warren Weaver, (1948) Science and Complexity, American Science, Volume 36, pages 536-544,

Wiki 2010 flash crash, https://en.wikipedia.org/wiki/2010_flash_crash

Michael J. Wooldridge. (1999) Intelligent Agents, chapter 1, pages 27–77. In Gerhard Weiss, Editor. Multiagent Systems – A Modern Approach to Distributed Artificial Intelligence. The MIT Press.

Co-evolving development, implementation and Operational SMS using a Digital Twin

Alastair Faulkner[1] and Mark Nicholson[2]

[1]Abbeymeade Limited, [2]University of York

Abstract Digital Twins (DT) are abstract, data-dependent and data-driven models. They are used to model the state of a physical component or system. This facilitates failure warnings and allows continuous improvement activities to be undertaken. As a result, DTs can be used as part of operational safety management. However, DTs also provide a tantalising opportunity to model whole systems before implementation, establishing a baseline model from which to identify safety and safety management issues relating to system realisation and operational safety management. They facilitate comparison between this model and reality as system realisation progresses. This study considers the use of DT in this role. Can the use of DTs mitigate issues that often delay the introduction to service? Can they be used early in the lifecycle to propose and check the credibility of 'approval-in-principle' documentation to ensure issues are discovered early rather than late in the project lifecycle? Benefits range from the application of change controls, impact assessment and the identification of interfaces and dependencies. This work sets out an approach to Systems and Safety Engineering using 'Engineered DTs' as part of safety assurance of current and future phases of a safety-critical project.

Keywords: Digital Twins, Data, Metadata, Metamodels

---

[1] email: alastair.faulkner@abbeymeade.co.uk
[2] email: mark.nicholson@york.ac.uk

© Alastair Faulkner and Mark Nicholson 2023.
Published by the Safety-Critical Systems Club. All Rights Reserved.

Architecting Safer Autonomous Aviation Systems

Jane Fenn[1], Mark Nicholson[2], Ganesh Pai[3] and Michael Wilkinson[1]

[1] BAE Systems, UK

[2] University of York, UK

[3] KBR / NASA Ames Research Center, USA

Abstract *The aviation literature gives relatively little guidance to practitioners about the specifics of architecting systems for safety, particularly the impact of architecture on allocating safety requirements, or the relative ease of system assurance resulting from system or subsystem level architectural choices. As an exemplar, this paper considers common architectural patterns used within traditional aviation systems and explores their safety and safety assurance implications when applied in the context of integrating artificial intelligence (AI) and machine learning (ML) based functionality. Considering safety as an architectural property, we discuss both the allocation of safety requirements and the architectural trade-offs involved early in the design lifecycle. This approach could be extended to other assured properties, similar to safety, such as security. We conclude with a discussion of the safety considerations that emerge in the context of candidate architectural patterns that have been proposed in the recent literature for enabling autonomous capabilities by integrating AI and ML. A recommendation is made for the generation of a property-driven architectural pattern catalogue.*

This work was co-authored by an employee of KBR, Inc. under Contract No. 80ARC020D0010 with the National Aeronautics and Space Administration. The United States Government retains and the publisher, by accepting the article for publication, acknowledges that the United States Government retains a non-exclusive, paid-up, irrevocable, worldwide license to reproduce, prepare derivative works, distribute copies to the public, and perform publicly and display publicly, or allow others to do so, for United States Government purposes. All other rights are reserved by the copyright owner

© BAE Systems, KBR, and University of York 2023.
Published by the Safety-Critical Systems Club. All Rights Reserved.

1 Introduction

Architecture and architecting have wide application in systems engineering, with architecture definition reportedly being one of the most often-used processes in model-based systems engineering (MBSE) (Cloutier and Bone 2015). Despite this, there is relatively little guidance available to practitioners on how to devise an architecture for a specific purpose, such as the incorporation of untrusted, but novel technologies, e.g., artificial intelligence (AI) and machine learning (ML), within a safety-critical system.

Within the aviation sector, and beyond, there has been considerable interest in the application of AI/ML to achieve autonomous operation. A key obstacle is the nexus of issues surrounding safety, including the assurance of safe autonomous operation, which is particularly acute when enabling technologies such as AI/ML may be non-deterministic and/or unpredictable at worst, and complex and opaque, at best.

In aviation, safety is the 'state in which risk is acceptable'. Practitioners in the domain recognise that choices made by system designers and implementers will have significant impact on the safety of a system, both on the allocation of requirements across the elements of the system design and also on how assurance of the implemented system is achieved. An architecture is commonly understood as the *organisation* or *structure* of a system in terms of its elements and their relationships (ISO/IEC/IEEE 2011, ISO/IEC/IEEE 2015).

This paper studies architectural structures that combine untrusted elements with trusted elements in such a way that the overall system can be considered safe. In this context, *safety* is an abstract property, which needs to be interpreted precisely in the context of the architectural structure proposed for a system.

Our paper is motivated by considerations of whether the way we currently architect aviation systems could help ensure that systems using AI/ML components are safe by design. We address these questions by assessing a set of existing and newly proposed standard architectural forms, *architectural patterns*, when AI/ML components are employed as part of the architecture[60]. We also give a reminder of the principles, objectives, and practices for architecture development in aviation as much of this knowledge is not explicit in current de facto standards (SAE 1996b), (SAE and EUROCAE 2010); rather it is implicit knowledge amongst the authors of those standards and guidance documents, and the practitioners in the domain.

---

[60] Though we refer to AI/ML broadly in this paper, we specifically consider neural networks (NNs), a particular approach to ML. We acknowledge that the terms *autonomy*, *artificial intelligence*, and *machine learning* have distinct meanings and implications, although they are often used interchangeably.

We hope this paper presents useful and interesting perspectives for many readers, but we anticipate the content being of particular interest to established safety engineering practitioners who are beginning to look at the issues with integrating AI/ML into their systems, and also to AI/ML development professionals who recognise that integrating their novel technology requires consideration of system safety. To appeal to as diverse an audience as possible, we avoid a presumption of a high level of prior knowledge and necessarily constrain our assessment to basic architectural patterns, deriving our observations from basic principles.

2 Background

2.1 Basic Concepts

Architecture concepts can be applied to any kind of system, at any level of system breakdown, and from any perspective (Wilkinson and Rabbets 2020).

Typically, certain perspectives are used to draw out types of architectural structure in both problem and solutions spaces, ranging from operational, through logical/functional to system/physical. Safety concerns may need to be addressed within each of these types of architectural structure.

A common assumption is that architects work 'top down': i.e., operational need informs the logical or functional structure that, in turn, determines the system or physical structure. In practice, this is often not the case. In this paper, we assume a new solution-space technology choice (AI/ML) has been selected. We then consider a set of logical or functional architectural patterns that could be employed and how AI/ML-based functions could be linked 'up' to the operational structure and 'down' to the system/physical structure.

In the aviation domain, an aircraft is a *product system* (INCOSE 2022) that exists within a broader system of systems. Each aircraft comprises interacting systems, themselves composed of subsystems and/or *line replacement units* (LRUs), which may have connections to resources such as sensors, effectors, electrical power, and cooling air. Within an aircraft, some avionic systems may employ a prescribed generic logical and physical architecture with defined interfaces between elements, e.g., *integrated modular avionics* (IMA), *full airborne capability environment* (FACE), and *common avionics architecture system* (CAAS).

A logical or functional architecture might indicate how functions and behaviour are allocated within a system element (e.g., LRU), paying attention to issues such as independence requirements. A physical architecture might show which components may be used. Where a system element contains diverse technologies, such as software and hardware in programmable devices, there will be distinct but related software and hardware architectures which together provide the overall architecture for an element of the system.

An AI/ML technology constitutes a specific type of programmable element; one whose behaviour may in general be non-deterministic, or at best be less transparent than conventional software and complex hardware. The precise nature of the challenges to be addressed architecturally depend on a multitude of factors, including the amenability of the element to analysis, its variability depending on learning algorithms and data, and its behaviour under anomalous conditions.

2.2 Architecture Principles

A system architecture can be used to establish an understanding of how the system is organised, e.g., how its constituent elements are structured, their boundaries, and the boundary to the environment. Additionally, it can serve to reconcile and realise competing requirements into a feasible basis to guide system design, development, and evolution. From a safety standpoint, the system architecture supports the principles and design decisions by which the emergent behaviour of the system (and that of its constituent elements) can be constrained to an acceptably safe operational state space.

Producing an architecture that is fit for purpose is a creative activity. In practice, architects use their experience to *hypothesise and test*: that is, define candidate architectural structures, informed by known requirements and constraints, and then assess whether those structures exhibit desired properties.

However, over the years architectural principles have emerged; for example, eleven principles or techniques for *fail-safe design* have been introduced in (FAA 1988, EASA 2021a) for the safety characteristics of aircraft. These include strategies such as redundancy or backup systems, warnings and error-tolerance, and the corresponding guidance suggests that a combination of two or more of the principles may usually be needed to achieve a fail-safe design. Wider dependability (reliability, availability, etc.) aspects also need to be considered as part of the architecting process.

A typical system architecture exhibits the following characteristics:

- the system organisation, boundaries, interfaces, and behaviours are refined to an appropriately low level [1];
- all established requirements (i.e., functional, performance, integrity, reliability, safety, security, robustness, and derived requirements, as well as those pertaining to interfaces, integration, and those stemming from physical, environmental, technological, and implementation constraints) are appropriately allocated;
- the stated system and safety requirements are satisfied;
- the following principles are leveraged as appropriate:
  - independence (e.g., in the safety-related features)

---

[1] i.e., An *item* level, based on the terminology used in civil aviation.

- diversity (functional, physical, design, data)
- layering, i.e., defence in depth
- avoidance, detection, and containment (e.g., through mechanisms such as redundancy, cross-checking, isolation, masking, etc.)

2.3 Developing an Architecture

There is ongoing work to provide guidance on safety assurance when integrating AI/ML into aviation systems (EASA 2021b, SAE and EUROCAE 2022) that outlines some candidate objectives, albeit from a process assurance standpoint.
In (SASWG 2022), specific objectives for architecture have been explicitly put forth as applicable in the context of autonomy, although process guidance has (intentionally) not been given.

For this paper, the architectural challenge is to accommodate elements incorporating currently untrusted technology (AI/ML) within an overall architecture that possesses adequate safety, defined in some way that is meaningful to stakeholders. In general this involves hypothesising an architecture and testing it, usually by means of analysis or argumentation of an appropriate kind, to ensure that it is safe.

Typically, an analyst would assess the safety impact of a range of failure modes using methods such as *Functional Hazard Assessment* (FHA), as in ARP 4761 (SAE 1996b). For software, limited guidance is available on the nature of failure modes to be considered.

A number of enhanced approaches have been proposed that are more tailored for complex software-intensive systems, such as *Software Hazard Analysis and Resolution in Design* (SHARD) (Pumfrey 1999), which employ a set of *guidewords* to establish software contribution to system safety. One such set is:

- *Omission*: The service is never delivered, i.e., there is no communication.
- *Commission*: A service is delivered when not required, i.e., there is an unexpected communication.
- *Early*: The service (communication) occurs earlier than intended. This may be absolute (i.e., early compared to a real-time deadline) or relative (early with respect to other events or communications in the system).
- *Late*: The service (communication) occurs later than intended. As with early, this may be absolute or relative.
- *Value*: The information (data) delivered has the wrong value.

These are useful concepts to consider when the system context is understood. However, analysts may require additional guidance on the interpretation of those guidewords and, indeed, additional guidewords may be appropriate when using AI/ML. We recommend this as future work, outside the scope of this paper.

Other techniques have also been proposed which take a broader perspective of the system in context of its operating environment (Leveson 2016). They similarly analyse deviations on expected performance at the system architecture level and assesses the potential safety impact.

Once safety requirements have been generated, defining an architecture usually involves function decomposition, trade-off analyses, consideration of design principles, as well as decomposition into lower levels of the system hierarchy. These activities should all be undertaken in the context of an assessment of the credibility of the design from a safety standpoint.

There are numerous potential approaches to aiding architectural selection such as the *Architecture Trade-off Analysis Method* (ATAM) (Kazman et al. 2000), and *Trade Trees* (NASA 2017). In general, these methodologies start by defining the scope or trade space, e.g., by defining the drivers for an architectural choice, such as tolerance to change, and various (prioritised) scenarios or configurations that characterise those drivers.

Then, the architectural options are identified together with a definition and prioritisation/weighting of the attributes of interest (e.g., reliability, safety, and security). Each option is analysed against each identified attribute and for each scenario, towards identifying (and selecting) optimised solutions.

Trade-offs are important drivers. For example, safety attributes may relate to not only the perceived ease in implementing safety requirements derived as a result of the architecture, but also the ease of generating the assurance evidence to demonstrate that the requirements have been met. Inspiration for hypothesising an architecture for a system may be drawn from previously successful architectures, captured as *architectural patterns*.

3 Architectural Patterns

The use of *patterns* has been recognised as a way of capturing good practice for many decades, initially in architecture, and subsequently, for system and software design, including successful safety approaches.

As software became routinely introduced into the development of systems whose behaviour could result in harm to humans, patterns of system-level safety protection architectures previously developed for hardware were updated. As software was introduced into aircraft systems, what were then new, and are now established, architectural patterns emerged. The initial version of the aerospace recommended practice for development of civil aircraft, ARP 4754 (SAE 1996a), lists the following patterns considered relevant to aerospace applications: (i) *partitioned design*; (ii) *dissimilar, independent designs implementing an aircraft-level function*; (iii) *dissimilar designs implementing an aircraft-level function*; (iv) *active-monitor parallel design*; and (v) *backup parallel design*. In much the same way, new architectures and architectural patterns are being proposed to integrate AI in general and ML in particular into safety-critical systems, especially in aviation.

What are the practical implications for selecting a system architecture with the move to integrating ML/AI? We revisit both the established and new architectural patterns, and investigate their suitability for use. We start by reviewing the simplest architecture: *single channel design* (Section 3.1.1). Then we examine the potential impacts when using two of the established patterns, active-monitor parallel design (Section 3.1.2) and backup parallel design (Section 3.1.3), with an AI/ML element, considering the contribution to system safety. We also examine how combining these established patterns (Section 3.1.4) relates to the newer proposed patterns (Section 3.2) and the considerations that emerge when including AI/ML based functions.

For what follows, we use *complex function* to mean an AI/ML-based complex function and will qualify the term when it is not clear from context.

3.1 Generic Architectural Patterns in Aircraft Systems

3.1.1 Single Channel Design

The single channel design (Fig. 1) is the simplest architecture possible for a system, where the complex element (originally software or complex hardware) inherits the totality of the safety requirements allocated to the function that it implements, in terms of failure probabilities (or failure rates) and assurance[2] requirements.

Failure rate requirements are expressed per average operational hour, while assurance requirements at a system level are determined (in civil aviation) from *Development Assurance Levels* (DALs)[3], i.e., levels A–E, with level A representing the highest criticality level and mapping to the most stringent requirements on process rigour.

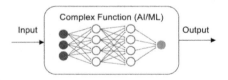

Fig. 1. Single channel design

---

[2] *Assurance* in this context means the approach used to manage systematic errors, which, in civil aviation, is usually associated with process rigour. In general, the term may be used slightly differently in other domains and in the associated standards/regulations.

[3] Functions are assigned so-called *function* DALs (FDALs). Upon decomposition and allocation to items, *item* DALs (IDALs) determine item-level assurance requirements. IDALs for software items are equivalent to (software) *design assurance levels* (DALs), the terminology used in the guidance documents for software assurance in aircraft system certification (RTCA and EUROCAE 2012).

Discussion. When the complex function in this pattern is implemented using AI/ML in general, and deep-learning in particular, it captures a so-called *end-to-end learning* architecture (Bojarski et al. 2016).

At present, there are neither broadly accepted techniques for determining failure rates for ML component failure modes that potentially contribute to system safety[4], nor is there an agreed approach to demonstrating achieved assurance. The probabilistic requirements for safety-critical systems are stringent, and it is not clear that current ML techniques can meet these requirements. For applications with minimal levels of safety requirements placed on the system, perhaps through their limited scope of use, the single channel architecture may present an acceptable residual risk. It is unlikely to be acceptable for moderate and higher levels of safety requirements allocated to a function.

3.1.2 Active-Monitor Parallel Design

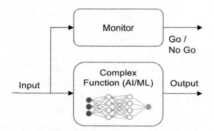

Fig. 2. Active-Monitor Parallel Design

Active-monitor parallel design (Fig. 2) is shown in the simplest, most generic form of architectural pattern. It requires a degree of interpretation for practical implementation: the monitor might be used either to disconnect the output from the complex function, or otherwise indicate the status of the output as *invalid* to downstream elements that consume the output. This pattern is predominantly about handling *erroneous function* or *malfunction* of the complex element, e.g., failure modes of *value* or *timing* that may have been derived, for example, using SHARD guidewords (see Section 2.3).

Discussion. In the simplest form, the monitor understands the transfer function between input and output of the complex element and uses the inputs to ascertain whether the output would be valid. Consider examples such as input range checking. In terms of allocating safety requirements, this pattern implicitly assumes that loss of function is of less concern. In terms of protection against erroneous function, subject to meeting availability requirements, a typical usage (SAE 2010) could be

---

[4] (Cluzeau et al. 2020) have claimed being able to determine ML failure rates in a safety-critical aviation application, although they have withheld material explanation of the underlying methodology from wider public scrutiny.

to allocate the highest assurance requirements (i.e., DAL A) to the monitor, to be implemented with simple, more readily verifiable technologies, while the more complex function then has to meet less onerous assurance requirements, e.g., those mapped to DAL C.

When an (AI/ML-based) complex function is used in this pattern, additional considerations are necessary around the monitoring function. Simplistic range checking may be required but may not be sufficient to determine whether the output of the complex function is safe in its system context. Sometimes AI/ML performance may be inadequate from a safety perspective within some parts of the operating domain. For example, the monitor needs to flag the outputs invalid at those parts. In this case, additional monitoring could be introduced at the risk of increased complexity, which may also make verification of the monitor more challenging. We will revisit these issues when we discuss the *runtime assurance* pattern (Section 3.2.1).

3.1.3 Backup Parallel Design

Backup parallel design (Fig. 3) is another simple form of architectural pattern that is helpful to ensure availability and to protect against a *loss of function* failure condition of the complex function.

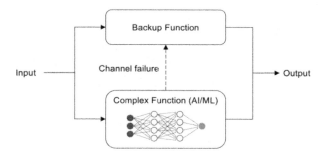

Fig. 3. Backup Parallel Design

Here, the obligation would be on the complex element itself to detect it is generating no output, or erroneous outputs, and self-suspend its function, as well as alert the backup function to take over. This suggests that it is possible to identify when the complex function has failed with a high degree of accuracy and certainty.

For safety requirements allocation, ARP4754A (SAE 2010) allows for DAL A requirements to be allocated such that, if sufficient independence could be shown, the primary (complex function) portion is implemented at DAL A and the backup portion at DAL C.

Discussion. When introducing AI/ML in the complex function, this allocation could be reversed so that the primary is allocated DAL C, with the backup allocated DAL A. Here, additional considerations emerge around both the self-test/self-diagnosis capability, and the balance between primary and backup functions. Self-diagnosis, and particularly detection of erroneous behaviour can become more complex than for traditional software. Hence, the patterns in the rest of this paper consider *channel error* rather than *channel failure*. Although the complex function may include non-ML elements such as pre- or post-processing within its boundary, to our knowledge limited evidence is available to conclude that those elements include self-diagnosis capabilities, for example, when AI/ML is used in perception pipelines.

The nature of the backup system requires specific considerations when using AI/ML. The pattern assumes an acceptable level of availability of the complex function, and one choice for the backup is functional equivalence to the complex element. Context-dependent decisions are necessary on whether the 'quality' of the backup function needs to be equivalent, or whether a 'limp home' *gracefully degraded* capability may be sufficient instead. The acceptability of such an approach will also be dependent on the anticipated balance of when the backup function will be the operational portion, based on how frequently the complex function reports as failed/invalid.

3.1.4 Combined Architectural Patterns

Industrial practice rapidly found it to be impractical to use simple patterns on their own, and currently it is more common to use a combination of patterns. We note that loss of function is typically more readily detected than erroneous function in conventional complex systems, and anticipate the same to be true for systems which will use AI/ML.

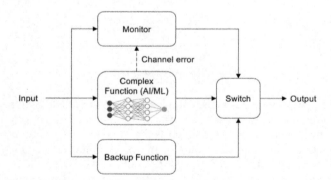

Fig. 4. A candidate combination pattern of active-monitor parallel design with backup parallel design

Based on the earlier discussions around the limited ability of AI/ML to detect its own failures, we now examine whether a combination of the active-monitor parallel design and the backup parallel design patterns could be useful to address the shortfalls of the individual patterns.

Several design choices are necessary when combining patterns to further address their individual shortcomings, so we illustrate an indicative implementation (Fig. 4) and consider a number of variants.

Discussion. For the pattern shown in Fig. 4, if the monitor determines that the complex function will not operate correctly, it switches to a conventionally assured backup function, assuming that an appropriate monitor can be constructed (see Section 3.2.1 for a more detailed discussion on this issue). As before, current practice would be to assign the overall pattern with safety targets in terms of failure rates and DALs. We assume that availability has been sufficiently addressed, and that erroneous function should be our main consideration.

When the pattern is used for conventional complex functions, the monitor and backup function inherit the same assurance requirements as the function allocated to the overall pattern. However, as previously mentioned, there are no broadly accepted techniques to determine the reliability of AI/ML-based complex functions, hence they need to be allocated lower target failure rates. In current practice, if the monitor and backup were allocated DAL A requirements for example, DAL C could be allocated to the (traditional software) complex function.

This allocation is not so straightforward for AI/ML-based complex functions. The combined architectural pattern of Fig. 4 requires that: (i) either the monitor understands the transfer function between inputs and outputs of the AI/ML-based complex function sufficiently to allow action to be taken when inputs are, for example, out of bounds; or (ii) the complex function can self-report erroneous function or loss of function. Where the monitor needs to switch to the backup, for instance in those regions of the *operational domain*[5] where the complex function performance has been determined to be inadequate, safety must be determined within the context of use.

If the complex function was introduced to improve capability, it is reasonable to infer that the backup function may not exactly replicate the ML-based function, and that the former's performance is an approximation of what the latter's output will be, either in the value domain, the timing domain, or both. In such cases, the switching behaviour at the appropriate portions of the input (i.e., the relevant regions of the operating domain), may itself be a safety property.

Moreover, it could be one that is defined during the development of the AI/ML-based complex function, with a lower level of rigour than that which would be necessary to support the overall allocated DAL for the monitor.

---

[5] We use 'operational domain' to mean the *Operating Domain Model*, or *Operational Design Domain* (*ODD*), which are the (descriptions of the) operational domains in which the AI/ML-based complex function is (to be) designed to properly function.

There are ramifications here for the cost of implementing this pattern. Current approaches assume that monitor development and assurance is feasible at a significantly lower cost and effort, relative to the complex function. The discussion above suggests that this is likely to be non-trivial. Also, as with the backup parallel design, the balance between the primary and backup functions needs to be considered for the pattern in Fig. 4. Next, we discuss whether the concerns, above, regarding the monitor could be addressed by using it differently.

Variants of Pattern Combinations. Using the monitor to observe the output of ML, rather than the inputs to the ML (Fig. 5) is one possible variation of the combined architectural pattern of Fig. 4.

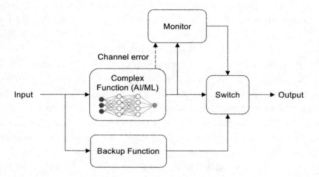

Fig. 5. Variant of the combined pattern of Fig. 4, with complex function *output* monitoring

In this configuration the monitor now judges whether the complex function output is valid against some defined criteria, relieving it of the requirement to know the transfer function implemented using AI/ML. However, the complexity is in defining the criteria for what constitutes *safe outputs*, particularly when the monitor has no understanding of the inputs that the complex function used to derive its outputs. Also, if the complex function is expected to process the inputs more rapidly than conventional means, for instance due to its optimisations and typical use of higher power and specialised computing hardware, there is a potential for a lack of synchronisation between the complex function output and what the monitor expects.

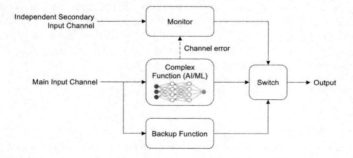

Fig. 6. Variant of the combined pattern of Fig. 4, with *independent channel input* monitoring

Another variation is to monitor the complex function inputs in addition to complex function outputs. In fact, this variant is the general configuration of the runtime assurance pattern discussed in more detail later (Section 3.2.1).

A third possible variant (Fig. 6) is to provide alternative inputs to the monitor. There is an independent sensing channel to observe environment conditions that are known a priori to degrade ML performance below acceptable levels. For instance, when using a complex function that was trained in high visibility conditions for vision-based perception in low light or low visibility conditions. In other words, the monitor checks for conformance to the operational domain defined for the complex function. This may provide additional confidence, providing the factors that cause poor ML performance are satisfactorily understood, with the rigour necessary to support the assurance requirements allocated to the monitor.

Model-centred assurance (Jha et al. 2020), a new alternative architecture proposed to enable autonomy, critiques this variant of the pattern. It asserts that 'perception functions of the primary system will surely be better resourced and more capable than those of the monitors' and concludes that the monitors should rather use the same input channels to construct a model of the environment.

The model-centred assurance architecture does not conform to the patterns considered in this paper. As such, we do not discuss it further here, leaving its assessment from an aviation safety standpoint to future work. Instead, we look at *other* architectural patterns from the literature that have been recently proposed for integrating AI/ML-based functionality into high criticality applications to enable autonomy.

3.2 Architectural Patterns for AI and ML

We consider the following four patterns in this section: *runtime assurance* (RTA), *value overriding*, *function modification*, and *input partitioning and selection*. Each pattern is briefly described, and their suitability for use in aviation is discussed, primarily from a safety standpoint.

3.2.1 Runtime Assurance

Runtime assurance (RTA) or its variations (Schierman et al. 2020, ASTM 2021) realised in a *system-level simplex* structure (Bak et al. 2009) is an architectural pattern for safety assurance of systems containing complex functions that cannot be approved to the requisite assurance level, for example as part of an aircraft system certification process that relies upon traditional development assurance.

The RTA pattern (Fig. 7) involves assured *runtime (safety) monitors*, receiving trusted inputs, that observe a complex, less-trusted or untrusted function. Upon detecting conditions that can violate safety, e.g., invalid inputs, deviant function outputs, or errant execution traces, the monitors trigger switching to one or more assured alternative functions to maintain a safe system state. To use the pattern, the RTA *wrapper* constituting the monitor(s), switch, and alternative function(s) must be assured to a higher level than the complex function.

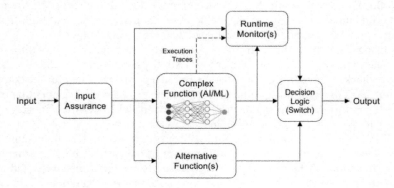

Fig. 7. Runtime assurance (RTA) pattern

There is compelling evidence that RTA works well in an aviation context (Burns et al. 2011). However, there are numerous considerations when using RTA for AI/ML-based functions, of which here we elaborate four.

Choice of the Complex Function Boundary. The AI/ML-based function boundary can have significant implications both on the assurance of the RTA scheme itself, and that of the integrated function (i.e., the AI/ML-based complex function secured within the RTA wrapper).

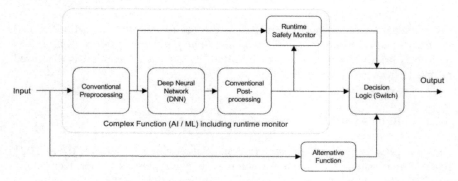

Fig. 8. RTA pattern including pre-/post-processing and the runtime monitor within the complex function boundary

One possible boundary for the complex function includes pre-/post-processing computation besides the ML model (Fig. 8). Pre-/post-processing using conventional (software) development techniques is typical in AI/ML development and may involve operations that are comparable to those that occur in the monitor, e.g., input/output range checking, or handling null values. Thus, it may be tempting to include the monitor within the function boundary (Fig. 8) and provide it with preprocessed inputs rather than the *raw* input. However doing so poses conflicts for allocating assurance requirements. Either the complex function and the monitor require the same level of assurance, which would violate the safety intent of RTA, where the monitor has a higher level of assurance than the function being monitored; or we must raise the level of assurance required for the complex function to that of the monitor, inducing a greater cost for generating the assurance evidence required.

Another possibility is to place only the ML model within the complex function boundary. In this case, either or both of pre- and post-processing will need the same level of assurance as the monitor.

Similarly, operations performed on the inputs necessary for high assurance could either be distributed across the monitors, complex functions, and the alternative functions(s), or be split off as a separate *signal conditioning*, or *input assurance* function (Fig. 7). In this case, other routine pre-processing, such as normalisation of data, would occur under lower assurance requirements within the complex function boundary.

Monitor Function Considerations. RTA requires a specification for the monitor that can be correctly implemented. This may be challenging in the context of AI/ML-based complex functions:

- Many monitors for detecting inputs that are not from the training distribution, i.e., *out-of-distribution* (OOD) inputs are themselves machine learnt. Similarly monitors for detecting out-of-operating domain inputs require assuming that the operating domain can be completely and comprehensively defined, which can be problematic for a perception function.
- Monitors to detect in-distribution inputs that could defeat the expected generalisation behaviour are difficult to specify and build because those inputs represent *surprises* that were previously unknown. Such inputs could be adversarial, or due to epistemic uncertainty about the operating domain.
- Depending on what the monitor is observing—e.g., raw inputs versus preprocessed inputs—value, timing, and synchronisation issues could potentially emerge that defeat the monitor, particularly while checking input/output validity.

Assurance Requirements Allocation. Where the RTA *wrapper* must itself be assured to the highest criticality level, i.e., DAL A, current development assurance processes require that:

1. the AI/ML function be assured to no lower than, in this case, DAL C. However, note that this pattern could be instantiated at different levels of abstraction, but assurance credit can only be taken *once* in the design. That

is, the reduction in assurance to DAL C is therefore only permissible where no credit has been taken at a higher level of abstraction, e.g., at the system level;
2. all failure conditions of the AI/ML-based function need to be correctly detected under all operating conditions;
3. assurance is needed that no AI/ML-based failure condition can cause monitor failure.

Although condition three may be achievable in principle through independence, partitioning, or isolation strategies it is unclear whether or not it is feasible to satisfy the former two requirements for AI/ML-based functions.

Configurations of RTA and Complex Functions. The RTA pattern (Fig. 7) belies complicated configurations that involve multiple monitors and alternative functions including hierarchies of the same. In the general case, the decision logic involves decision-making under uncertainty which may, itself, be a complex (possibly AI/ML-based) function, thus making assurance of the RTA scheme at least as difficult as the assurance of the complex function. If the decision logic can be precisely specified, the safety assurance situation is more promising: it may be feasible to formally verify an AI/ML-based implementation of the same (Katz et al. 2017).

Apart from the monitor and backup functions, the complex function may itself be achieved using multiple simpler models rather than a monolithic model, as is common in *ensemble learning*. This might offer additional options for monitoring (Fig. 9) The individual models of the complex function may be simple and therefore monitoring those may also be simple. More analysis is needed however, to determine whether this simplifies or complicates monitoring overall, particularly due to issues of monitor consistency, timing, throughput, and synchronisation, for example.

As mentioned earlier (Section 3.1.4) the RTA pattern is, in principle, one variant of the combination of the active-monitor parallel design and backup parallel design patterns. Thus, many of the considerations applying to those simpler patterns also emerge here.

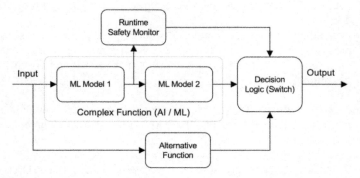

Fig. 9. RTA pattern including multiple ML models within the complex function

For instance, how often the alternative function is required to be operational, versus how frequently complex function failure is accepted; and the safety impact of the switching behaviour with respect to portions of the input domain where the complex function underperforms, and the corresponding timing requirements.

3.2.2 Value Overriding

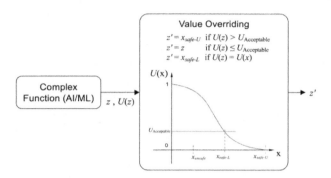

Fig. 10. Value overriding pattern (Non-adaptive)

In (Groß et al. 2022), four patterns have been proposed in the context of self-driving road vehicles for 'handling runtime uncertainty in perception', namely *uncertainty supervisor*, *safety margin selector*, *adaptive uncertainty supervisor*, and *adaptive safety margin selector*. Each of the proposed patterns may be applied to processing sensed data that have an associated uncertainty, e.g., an AI/ML model-based estimate of a quantity such as the coefficient of friction between vehicle tyres and the road surface, before they are presented as input to any subsequent computation.

The pattern to be used depends on how the data and its associated uncertainty are presented. The uncertainty supervisor pattern applies to point values, where a worst-case value replaces the input received when the uncertainty for that input exceeds a predetermined acceptable threshold. The safety margin selector pattern may be applied for input presented as a distribution, where the value replacing the input is, again, chosen using a predetermined threshold for acceptable uncertainty.

Effectively, these two patterns are variations of a common pattern of *value overriding* (Fig. 10) where the given data are overridden with *safe* values when their associated uncertainty exceeds some acceptable threshold. The *adaptive* variations of value overriding (i.e., adaptive uncertainty supervisor, and adaptive safety margin selector) vary the threshold for acceptable uncertainty itself, based on an additional input (not shown in Fig. 10) representing the (risk of the) operating situation.

Discussion. Applying the value overriding pattern induces additional requirements, not readily evident in (Groß et al. 2022):

- *worst-case safe values exist that can be independently established*: this requirement constrains the scope of the perception problem for which the AI/ML function is being used, and thus the kind of responses to which the pattern may be applied. In principle, this requirement may be achievable for quantities for which ground-truth worst-case reference values can be separately measured, determined through mathematical modelling, or using simulation.
- *reference uncertainty distributions used to select safe replacement values are valid and accurate*: as with the previous requirement, this requirement also could be met, in principle, through measurement, modelling, and simulation provided that there exists a ground-truth basis to the responses for which an uncertainty distribution is being determined.
- *uncertainty estimates produced by the AI/ML function can be trusted*: satisfying this constraint will be problematic when incorrect responses are produced with high confidence (equivalently, low uncertainty), e.g., in the presence of adversarial inputs to ML models, or unexpected inputs drawn from distributions differing from the training distribution.
- *operating situations are correctly determined*: although it may be possible in specific cases to use existing sensors and techniques for establishing the operating situation, in general this is itself the perception problem. Effectively, a circularity of requirements emerges where assured perception is needed to provide assurance of quantities themselves determined from perception.

The preceding requirements are not a comprehensive set and more analysis is needed to determine the additional requirements needed to usefully apply the value overriding pattern. For example, in conjunction with other architectural patterns, for different system configurations, and for different function criticalities.

By design, the value overriding pattern can modify the allowable safety margin in the responses of AI/ML-based functions. The justification for this choice is that (Groß et al. 2022):

> …runtime estimation and handling of uncertainties is necessary to overcome worst-case approximations that would lead to unacceptable utility/performance, especially if the situation context indicates a low risk situation.

From an applicability in aviation standpoint, the principle underlying this pattern—that greater uncertainty in the responses from an AI/ML function is tolerable for performance gains in low(er) risk operating situations—violates the intent of the fail-safe design concept required by aircraft airworthiness regulations (FAA 1988, EASA 2021a). More specifically it violates *principle 10: margins or factors of safety to allow for any undefined or unforeseeable adverse conditions*. The rationale is that worst-case estimates for a quantity are rather a worst-case *lower bound*. Thus, reducing that safety margin introduces the additional burden of demonstrating that operating at any margin down to the alternative worst-case lower bound does not exhibit any unintended behaviour with unacceptable safety impact, under all foreseeable operating conditions.

Furthermore, in aviation systems of systems (e.g., air traffic management (ATM), air navigation services (ANS), airport operations) safety margin reductions in favour of performance improvement can accumulate over time leading to *practical drift*, i.e., where safety performance is presumed safe but is in fact in an unforeseen, and appreciably higher-risk region than the approved baseline (ICAO 2018). As such, applying this pattern in its current form, especially for functions assigned higher DALs is unlikely to be acceptable.

3.2.3 Function Modification

Deep neural networks (DNNs) used for object detection as part of a vision-based perception system typically process sequences of input images and produce *bounding boxes* that spatially localise and highlight the detected objects of interest on each image. In use cases such as collision avoidance, the safety contribution of object detection to system hazards can be characterised by *false negative detections* (i.e., an object posing a collision hazard exists in the image, but the DNN does not recognise it) and *inaccurate localisations* (i.e., an object posing a collision hazard that exists in the image is correctly recognised, but the bounding box produced either partially covers it, or does not cover it).

For positive detections, *Intersection over Union* (IoU) is a frequently used metric of bounding box estimation performance, measuring the extent of overlap of the ground-truth and the predicted bounding boxes to quantify localisation accuracy. Thus, evidence from development that IoU is perfect (or nearly perfect) for all test input images not seen in training is desirable for safety assurance that the predicted bounding boxes will be accurate in deployment.

Safety post-processing (Schuster et al. 2022) has been put forth as a solution for this purpose in the automotive systems domain, for collision avoidance. Specifically, after conventional post-processing of the 2D, rectangular, and axis-aligned bounding boxes that the DNN produces on test images, safety post-processing scales them by an enlargement factor proportional to the IoU that was established during training. The possible range of values of the enlargement factor are mathematically proved to be the smallest required to guarantee that the enlarged bounding boxes will always contain the ground-truth object for detections on all input images, for all values of IoU obtained in training.

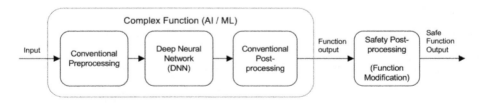

Fig. 11. Safety post-processing architecture, adapted from (Schuster et al. 2022), as an instance of the function modification architectural pattern

In general, this architecture could be seen as an instance of a pattern of (assured) *function modification* (Fig. 11) where the modified function outputs are (to be) guaranteed to meet the relevant safety constraints on the AI/ML function.

Discussion. The IoU metric admits values in the interval [0, 1], whereas the theorem that relates the enlargement factor to IoU (guaranteeing that ground-truth objects are always included in the scaled bounding boxes after safety post-processing) applies to the half-open interval (0, 1] (Schuster et al. 2022).

Since IoU = 0 for false positive detections, and neither true nor false negative detections produce bounding boxes, safety post-processing can only apply to the bounding boxes of true positive detections, and other mechanisms are required to minimise and mitigate false detections.[6]

The latter may occur due to inadequate generalisation of learnt behaviour to nominal, out-of-sample, in-distribution inputs[7] (including edge and corner cases), insufficient robustness to adversarial, out-of-sample, in-distribution inputs, and/or OOD inputs that may be benign or adversarial.

In general, assurance that the learnt behaviour of the AI/ML function is robust and as expected over the required domain of in-distribution inputs is necessary for function modification to be usefully applicable. Additionally, applying the function modification pattern for perception, as shown in Fig. 11, in a single channel configuration (see Section 3.1.1) may not be sufficient for safety assurance. It requires combination with other architectural patterns. For example:

- a self-checking pair that includes diversity in the sensing and object detection functions, possibly in separate channels, to recognize false detections from DNN-based perception, with voting and/or sensor fusion to resolve disagreement in object detections;
- the active-monitor parallel design pattern that includes monitoring to detect out-of-distribution (OOD) inputs, coupled with OOD input handling.

3.2.4 Input Partitioning and Selection

The *input partitioning and selection* architectural pattern (Fig. 12) contains two or more channels, each performing the same function, but on different partitions of the input domain of the function. One or more of the channels may include AI/ML, but at least one channel does not.

---

[6] Here, although false positive detections do not have an immediate worst-case safety impact (since there are no obstacles posing a collision hazard), non-detection of a false positive either may be a precursor to a hazard—e.g., taking a corrective action when not required may itself be a hazard—or may lead to an effect of lower (but not insignificant) safety criticality, e.g., increased pilot or operator workload in an aviation context.

[7] Operational input data that are from the same distribution as the training and testing data (i.e., in-distribution), which were not sampled for inclusion into the training and testing required during the learning process (i.e., out-of-sample).

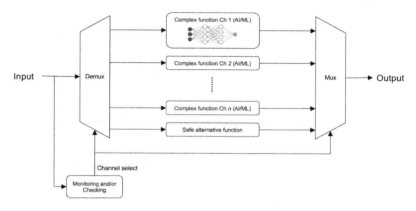

Fig. 12. Input partitioning and selection architectural pattern

Conceptually, this pattern may be understood as a combination of a *demultiplexer* of the function input to different channels, and a *multiplexer* of different channels to the function output. The channel to be selected relies on monitoring of predefined conditions (e.g., whether the inputs are within certain bounds) and OOD inputs, or checking of predefined properties (e.g., whether the expected output is produced for a given input).

The 'hybrid' architecture proposed in (Damour et al. 2021) is an instance of this pattern containing 2 channels. The primary channel comprises a DNN-based implementation of the next-generation airborne collision avoidance system for unmanned aircraft (ACAS-Xu), which produces collision avoidance advisories for the given inputs. In fact, *multiple* DNN models are used to implement this complex function (an option previously considered in Section 3.2.1). The second channel is a lookup table (LUT)-based 'safety net' representing an alternative implementation of ACAS-Xu. This LUT-based channel is meant to operate on that portion of the input space for ACAS-Xu, where the DNN-based implementations are known to perform poorly, as established during their machine learning development lifecycle.

Discussion. In its simplest incarnation, all channels are required to cover the complete input space for the function allocated to the pattern, and each channel operates only on a specific portion of that input space. Thus, at any given time, there is exactly one channel that can produce the required output for a given input, when that channel is operational. Loss of any channel thus leads to a loss of the function on the corresponding portion of the input.

For high criticality functions, high assurance is needed that: (i) the monitor/checking function invokes the safe alternative/backup function *only* in those portions of the operating domain where the primary channels and the backup are known to diverge; (ii) for the remaining portion of the operating domain, inputs (outputs) are correctly routed to (from) the appropriate channel; (iii) the safe backup is correct against a validated specification of the function allocated to the pattern; and (iv) the primary channel implemented using AI/ML is correct against the validated specification of the function for those portions of the operating domain where its responses are consistent with the safe backup.

There are numerous implications of the above assurance requirements. The first two conditions relate to correctness of the monitor/checking function. They additionally require the safe backup and the primary channel to be consistent in their outputs for some common portion of the operating domain, and that consistency be shown with high assurance. In the hybrid architecture (Damour et al. 2021), formal verification is used to establish this consistency property, although in general (e.g., for a perception function), this may be challenging to achieve or demonstrate.

The implication of condition three, above, is that the safe backup is itself a complex function albeit not one implemented using AI/ML, but using a conventional approach instead. In other words, it could in principle serve as a primary channel over the entire operating domain of the intended function. This necessarily requires complete functional equivalence of the AI/ML-based complex function and the safe backup, even if each is only being used on specific portions of the operating domain. Again, in general, this may be challenging to achieve, and might raise the (legitimate) question of needing to use an AI/ML implementation in the first place.

For the hybrid architecture, the answer lies in the appreciable memory and power savings from using DNNs versus LUTs. As such, this architecture may be largely beneficial for achieving objectives other than safety: indeed, at a higher level of abstraction, this pattern is effectively an instance of single channel design (Section 3.1.1). Thus, all the channels and the monitor/checking function inherit the totality of the assurance requirements for the function that is allocated to the pattern.

That, in turn, affects condition four above. Specifically, there is no relief in the level of assurance for the primary channel as in other architectural patterns involving a safe alternative function. In the hybrid architecture, assurance of the DNNs is expected to be shown using *learning assurance*, a new process being defined in (SAE and EUROCAE 2022) and first conceived in (EASA 2021b). The latter does not yet support high-criticality applications, while it remains to be seen whether regulators will endorse the former as an acceptable means of compliance to airworthiness regulations.

As such, for functions assigned higher DALs, the input partitioning and selection pattern will likely to be combined with other patterns such as RTA, backup parallel design, or *triple modular redundancy* (TMR) and the combination analysed together, to establish their suitability for use.

4 Concluding Remarks

Although others have assessed both hardware and software architectural patterns from a safety standpoint (Armoush et al. 2009, Hammett 2002), the integration of AI/ML was not a consideration. In (Armoush et al. 2009), the main safety focus was on achievement of reliability targets, related to so-called *safety integrity levels*

(SILs), an orthogonal concept to development assurance levels (DALs) as we have considered in this paper.

The architecting problem when using ML/AI components has led to several variants of the RTA and system-level simplex architectural patterns—e.g., *certified control* (Jackson et al. 2021), and synergistic redundancy (Bansal 2022)—as well as proposals for novel architectures not conforming to previously established patterns or their combinations, e.g., *model-centred assurance* (Jha et al. 2020). Other architectural patterns established for functional safety have also been analysed from the standpoint of SIL allocation (Koopman 2021). However, the application domain for all of the above is autonomous road-vehicles.

Candidate architectures and patterns need to be assessed for their suitability in aviation applications so that guidance can be provided on how to justify credibility of the chosen form. This assessment will have to take into account compatibility with existing architectures, e.g., integrated modular avionics (IMA) used in aircraft systems. We have taken the first steps in this paper to identify and collate architectural patterns, both established and newly proposed, for including AI/ML in an aviation context, and assessed their potential suitability for use when viewed through the lens of safety assurance.

The DAL paradigm for sufficiency of safety assurance largely influences the choice of an architecture in an aviation system (in particular for aircraft systems), under the current certification regime. That is, DALs modulate how much assurance is required in proportion to the safety criticality of a function. They translate a level of assurance (e.g., Levels A - E) to the extent of development rigour necessary, which is codified in terms of process objectives. The higher the DAL, the greater the development rigour needed, and more evidence needs to be produced. Guidance does not currently exist for the nature of the evidence that will be needed to provide appropriate confidence that proposed architectures including ML/AI are fit for purpose. Although process assurance guidance for AI/ML-based products is being developed (SAE and EUROCAE 2022), the guidance it contains on architecture and architecting is largely implicit.[8]

The discussion earlier in this paper suggested that the prevailing architectural patterns (e.g., from ARP 4754) remain valid in principle when used with AI/ML-based complex functions. However, in practice they will need to be modified to address the nature of the failure modes of ML components. Such adjustments are also likely to be needed by other architectural patterns used in aviation (Hammett 2002) that we have not analysed in this paper. Although new architectural patterns have been proposed for integrating AI/ML, adopting them for use in aviation is far from straightforward. Indeed, in their current form, some may be unsuitable for use for high-criticality functions when confronted with the stringency of the associated assurance requirements. Moreover, the complexity required of components such as monitors may mean that the effort and cost of employing such architectures is increased.

Our discussions lead us to the following recommendations:

---

[8] Two of the authors are members of the standardisation committee that is formulating the guidance in (SAE and EUROCAE 2022), and the associated technical exchanges have had a part in motivating us to craft this paper.

1. Generation of a catalogue of architectural patterns and, potentially, combinations thereof for safety of systems containing untrusted technology—for systems in general, and aviation systems in particular.
2. An assessment framework by which the suitability of a proposed architectural pattern can be assessed for credibility during system development. For example, work is required to identify proportionate assurance requirements over the architectures and to determine whether additional guidewords are required for safety assessment when utilising AI/ML.
3. Development of assurance or confidence case patterns associated with the use of the architectural patterns in the catalogue.

Finally, this work is at an early stage. We welcome feedback on both the usefulness of these recommendations, and the ways to fulfil them.

Acknowledgements Ganesh Pai contributed to this work under support from the System-wide Safety (SWS) project under the Airspace Operations and Safety Program of the NASA Aeronautics Research Mission Directorate (ARMD).

Disclaimer The opinions, findings, recommendations, and conclusions expressed are those of the authors and do not represent the official views or policies of KBR, Inc., the National Aeronautics and Space Administration, and United States Government.

References

Armoush, A., Beckschulze, E., Kowalewski, S. (2009) Safety Assessment of Design Patterns for Safety-Critical Embedded Systems, In: 35th Euromicro Conference on Software Engineering and Advanced Applications, pp. 523-527. DOI: 10.1109/SEAA.2009.12

ASTM (2021). F3269-21 Standard Practice for Methods to Safely Bound Behaviour of Aircraft Systems Containing Complex Functions Using Run-Time Assurance. ASTM International.

Bak, S., Chivukula, D., Adekunle, O., Sun, M., Caccamo, M., Sha, L. (2009) The System-Level Simplex Architecture for Improved Real-Time Embedded System Safety. In: 2009 15th IEEE Real-Time and Embedded Technology and Applications Symposium, San Francisco, CA. pp. 99–107. DOI: 10.1109/RTAS.2009.20

Bansal, A., Yu, S., Kim, H., Li B., Hovakimyan, N., Caccamo, M., Sha, L. (2022). Synergistic Redundancy: Towards Verifiable Safety for Autonomous Vehicles. arXiv:2209.01710 [cs.RO]

Bojarski, M., del Testa, D., Dworakowski, D., Firner, B., Flepp, B., Goyal, P., Jackal, L., Monfort, M., Muller, U., Zhang, J., Zhang, X., Zhao, J., Xieba, K. (2016). End to End Learning for Self-Driving Cars. arXiv:1604.07316v1 [cs.CV]

Burns, A., Harper, D., Barfield, A., Whitcomb, S., Jurusik, B. (2011) Auto GCAS for Analog Flight Control System. In: 2011 IEEE/AIAA 30th Digital Avionics Systems Conference. DASC 2011, Seattle, WA, pp. 8C5-1-8C5-11. DOI: 10.1109/DASC.2011.6096148.

Cloutier, R., Bone, M. (2015) MBSE Survey, In: MBSE Workshop at the INCOSE International Workshop, Torrance, CA, USA. (Online): https://tinyurl.com/incose-mbse-survey-2015 (Accessed 7 November 2022)

Cluzeau, J.M., Henriquel, X., Rebender, G., Soudain, G., van Dijk, L., Gronskiy, A., Haber, D., Perret-Gentil, C., Polak, R (2020) Concepts of Design Assurance for Neural Networks (Co-DANN). Public Report Extract ver. 1, EASA AI Task Force, and Daedalean AG. (Online): https://www.easa.europa.eu/en/downloads/112151/en (Accessed 28 October 2022)

Damour, M., de Grancey, F., Gabreau, C., Gauffriau, A., Ginestet JB., Hervieux, J., Huraux, T., Pagetti, C., Ponsolle, L., Clavière, A. (2021). Towards Certification of a Reduced Footprint ACAS-Xu System: A Hybrid ML-Based Solution. In: Habli, I., Sujan, M., Bitsch, F. (eds) Computer Safety, Reliability, and Security. SAFECOMP 2021. LNCS 12852. Springer. DOI: 10.1007/978-3-030-83903-1_3

EASA (2021a) Certification Specifications and Acceptable Means of Compliance for Large Aeroplanes. CS 25 Amendment 27. European Union Aviation Safety Agency.

EASA (2021b) Concept Paper: First Usable Guidance for Level 1 Machine Learning Applications. Issue 01. European Union Aviation Safety Agency.

FAA ANM-110 (1988) System Design and Analysis. Advisory Circular AC 25.1309-1A. Federal Aviation Administration, US Department of Transportation.

FAA AIR-120 (2013) Integrated Modular Avionics Development. Verification, Integration and Approval using RTCA/DO-297 and Technical Standard Order C153. Advisory Circular AC 20-170 Including Change 1. Federal Aviation Administration, US Dept. of Transportation.

Groß, J., Adler, R., Kläs, M., Reich, J., Jöckel, L., Gansch, R. (2022) Architectural Patterns for Handling Runtime Uncertainty of Data-Driven Models in Safety-Critical Perception. In: Trapp, M., Saglietti, F., Spisländer, M., Bitsch, F. (eds) Computer Safety, Reliability, and Security, SAFECOMP 2022, LNCS 13414, Springer. DOI: 10.1007/978-3-031-14835-4_19

Hammett, R. (2002) Design by Extrapolation: An Evaluation of Fault Tolerant Avionics. In: IEEE Aerospace and Electronic Systems Magazine, 17(4), IEEE pp. 17-25. DOI: 10.1109/62.995184.

ICAO (2018) Doc 9859, Safety Management Manual (4th ed.) International Civil Aviation Organisation.

ISO/IEC/IEEE (2015) International Standard 15288-2015 System and Software Engineering – System Life Cycle Process.

ISO/IEC/IEEE (2011) International Standard 42010:2011, Systems and Software Engineering – Architecture Description.

Jackson, D.; Richmond, V.; Wang, M.; Chow, J.; Guajardo, U.; Kong, S.; Campos, S.; Litt, G.; and Arechiga, N. (2021) Certified Control: An Architecture for Verifiable Safety of Autonomous Vehicles. arXiv:2104.06178 [cs.RO]

Jha, S., Rushby, J., Shankar, N. (2020) Model-Centered Assurance for Autonomous Systems. In: Casimiro, A., Ortmeier, F., Bitsch, F., Ferreira, P. (eds) Computer Safety, Reliability, and Security. SAFECOMP 2020. LNCS 12234. Springer. DOI: 10.1007/978-3-030-54549-9_15

Katz, G., Barrett, C., Dill, D., Julian, K., Kochenderfer, M. (2017) Reluplex: An Efficient SMT Solver for Verifying Deep Neural Networks. In: Majumdar, R., Kunčak, V. (eds) Computer Aided Verification. CAV 2017. LNCS 10426. Springer. DOI: 10.1007/978-3-319-63387-9_5

Kazman. R., Klein. M., and Clements, P. (2000) ATAM: Method for Architecture Evaluation. Technical Report CMU/SEI-2000-TR-004. Software Engineering Institute, Carnegie Mellon University, Pittsburgh, PA.

Koopman, P. (2021) Safety Architecture Patterns. Lecture Notes of Electrical and Computer Engineering Graduate Course (18-642) on Embedded System Software Engineering. Carnegie Mellon University, Pittsburgh, PA (Online): https://archive.org/details/35-safety-arch-patterns (Accessed 7 November 2022)

Leveson, N., (2016) Engineering a Safer World - Systems Thinking Applied to Safety. MIT Press.

NASA (2017) Systems Engineering Handbook. NASA/SP-2016-6105 Rev. 2. National Aeronautics and Space Administration, Washington, D.C.

Pumfrey, D. (1999) The Principled Design of Computer System Safety Analyses. PhD Thesis, University of York, Department of Computer Science, York, UK.

RTCA and EUROCAE (2012) DO-178C/ED-12C, Software Considerations in Airborne Systems and Equipment Certification, RTCA Inc. and EUROCAE

SAE Systems Integration Requirements Task Group (1996a) Certification Considerations for Highly-Integrated or Complex Aircraft Systems. Aerospace Recommended Practice ARP 4754, SAE International.

SAE G-34 and EUROCAE WG-114 (2022) Process Standard for Development and Certification / Approval of Aeronautical Safety-Related Products Implementing AI, AS6983 (Work in Progress). SAE International and EUROCAE.

SAE S-18 Aircraft and System Development and Safety Assessment Committee (1996b) Guidelines and Methods for Conducting the Safety Assessment Process on Civil Airborne Systems and Equipment. Aerospace Recommended Practice ARP4761, SAE International.

SAE S-18 and EUROCAE WG-63 (2010) Guidelines for Development of Civil Aircraft and Systems. Aerospace Recommended Practice ARP4754A, SAE International / EUROCAE Document ED-79A, EUROCAE.

Safety of Autonomous Systems Working Group (SASWG) (2022). Safety Assurance Objectives for Autonomous Systems version 3.0 SCSC-153B. Safety Critical Systems Club.

Schierman, J., DeVore, M., Richards, N., Clark, M. (2020) Runtime Assurance for Autonomous Aerospace Systems. Journal of Guidance, Control, and Dynamics 43(12), pp. 2205-2217. AIAA. DOI: 10.2514/1.G004862

Schuster, T., Seferis, E., Burton, S., Cheng, CH. (2022). Formally Compensating Performance Limitations for Imprecise 2D Object Detection. In: Trapp, M., Saglietti, F., Spisländer, M., Bitsch, F. (eds), Computer Safety, Reliability, and Security SAFECOMP 2022, LNCS 13414. Springer. DOI: 10.1007/978-3-031-14835-4_18

INCOSE (2022) Systems Engineering Body of Knowledge (SEBoK) ver. 2.6, Part 4 Product Systems Engineering (Online) https://www.sebokwiki.org/wiki/Product_Systems_Engineering (Released 20 May 2022, Accessed 30 October 2022)

Wilkinson, M. and Rabbets, T. (2020) Don't Panic - The Absolute Beginner's Guide to Architecture and Architecting. INCOSE UK, Ilminster.

Grasping the Chalice – The Quest for the "Holy Grail" of Drone Operations

Paul Hampton

CGI

London, UK

Abstract *The use of drones has been hailed for many years as the next multi-billion-dollar industry and technological disruptor. While swarms of drones do not yet blacken our skies, there is evidence that the use of drones is starting to gather pace; Royal Mail post and Covid-19 medical deliveries, the use of drones in the war in Ukraine, and even, the aerial lightshow at the Queen's Platinum Jubilee Party, show that drones are becoming more prevalent in the public consciousness. However, the main barrier to much more prolific drone use is the regulatory challenges in flying drones Beyond Visual Line of Sight (BVLOS). BVLOS is something of a "Holy Grail" for drone operations as it unlocks an immense number of novel business opportunities and means to radically improve efficiency and safety of existing operations. The UK is leading the way in BVLOS operations, with much continued government funding, but although regulations and means of compliance are starting to emerge for aspects of general BVLOS operations, the path to routine commercial operations remains challenging. The risks of BVLOS flights are currently managed by regulators on a case-by-case basis and those operations that have been approved are generally flying in remote geographies away from populated areas, or in segregated airspace specially provisioned for drone flights. This paper explores the challenges and practical realities for BVLOS operations in higher-risk locations, such as urban areas and in controlled and unsegregated airspace, and assesses how stakeholders across the entire ecosystem, from operators through to regulators, are progressing towards making these types of operation more routinely certifiable.*

© Paul Hampton 2023.
Published by the Safety-Critical Systems Club. All Rights Reserved

1 Introduction

1.1 Terminology

While "drone" is perhaps the most commonly used term by the public and media, more formal standards and regulations refer to drones as Unmanned Aerial Vehicles (UAV) or Remotely Piloted Aircraft (RPA). The aircraft along with its supporting equipment and infrastructure required for operation (control station, remote link etc.) are then termed the Unmanned Aerial System (UAS) and Remotely Piloted Aircraft System (RPAS) respectively. UAS and RPAS terms are not however interchangeable as they encompass different sets of capability. The non-gender-neutral nature of these UAV/UAS definitions is also leading to the use of "uncrewed" rather than "unmanned". Although many documents will still refer to the former, there is strong precedence being set with the US Government already making changes to their documents (Washington Times 2022).

UAVs, RPAS and BVLOS operations were described by Hampton, Pugh and Ball (2020). That material is not reproduced here, but some of the key concepts described there are summarised and refined as follows.

An RPAS is defined by the ICAO Manual of Remotely Pilot Aircraft Systems (ICAO 2015) as:

> "A remotely piloted aircraft, its associated remote pilot station(s), the required command and control links and any other components as specified in the type design."

A distinguishing feature of an RPAS system is that it is an aircraft operated without the possibility of direct human intervention from within or on the aircraft. An RPAS system includes the airframe, propulsion unit, flight controls, health monitoring systems, data communications, electrical system, navigation system, sensors and any other component on-board or attached to the aircraft.

The RPAS has a pilot who is located remotely to the aircraft who controls and manages the aircraft during the total duration of the flight (take off, air manoeuvres and landing).

A UAS is a broader term as it also encompasses fully automatic and autonomous vehicles as well as those used to support remote piloting operations. An "autonomous aircraft" is defined by ICAO (ICAO 2015) as:

> "An unmanned aircraft that does not allow pilot intervention in the management of the flight"

The most recent guidance from the UK's aviation regulator – the Civil Aviation Authority (CAA 2022) – indicates that the CAA make a strict demarcation between *fully automatic* and *autonomous* systems. They have introduced a 5-level scheme categorising increasing levels of automation with the highest, similar to that introduced for road vehicles (SAE 2021). Level 5 – "full automation" is defined by the CAA as:

> "there is no human involvement in the operation and human interaction is limited to providing high-level operational directives and observing resulting outcomes. No human intervention is possible as the operation outcomes are entirely within the scope of the machine".

This might, to some, be an equally good definition of an *autonomous* system but the CAA seem to reserve the definition of an autonomous systems to those that exercise *non-deterministic* behaviour. This is perhaps an unfortunate choice of association as some Artificial Intelligence (AI) systems that have neural networks trained using Machine Learning (ML) techniques are entirely deterministic; the same inputs fed into the network will always yield the same outcomes. It is perhaps the explainability of the resulting behaviour that is the real issue to address.

From these definitions, an RPAS is a type of UAS, but there are some UAS[1] that are not classed as RPAS. This is because some UAS will be fully automatic/autonomous and have no remote piloting capability.

Some definitions of a UAV (such as the Wikipedia definition as of Dec 2022) assert that the UAV is not passenger bearing, yet this is not precluded by the ICAO definitions and there is a real prospect of remotely piloted urban air taxis in coming years, so passengers should be considered possible, and simply treated as another form of payload for a UAV.

One other form of remotely piloted aircraft system is where the aircraft also has a crew on board. For example, one future concept involves long haul air transport where a remote pilot acts as relief crew during quieter cruise phases so fewer pilots are required on the aircraft. This type of remote piloting is not precisely catered for by the RPAS definition but is entirely relevant to the discussion and, in fact, could arguably see real operations sooner than some of the other entirely uncrewed solutions. Other more exotic forms of future piloting may also ill-suit the term:

- A UAV that detaches from, but is still controlled from a piloted vehicle;
- A UAV piloting another UAV;
- A swarm of UAV's making collective piloting decisions for the entire group.

Perhaps a new term such as "Proxy Piloted Aircraft System" (PPAS) would be a cure to all these difficulties.

The following diagram is provided to help illustrate these differences and cover all scenarios. It is based on five main considerations: whether the aircraft is:

---

[1] To avoid awkward expressions such as "UASs" or "UAS's", for the purposes of this paper, the terms RPAS and UAS import both singular and plural meaning.

1. crewed (C)
2. uncrewed (U)
3. carries passengers (P)
4. autonomous (A)
5. remotely piloted (R)

which gives rise to the name of the diagram: CUPAR.

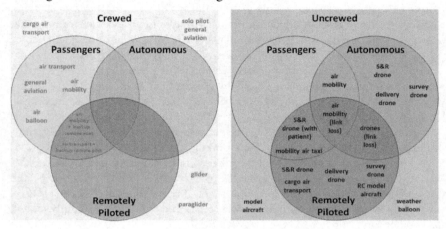

Fig. 1. CUPAR Diagram

In the Uncrewed side of the figure, UAV relates to everything in the Remotely Piloted and Autonomous circles. RPAS relates only to Remotely Piloted section that intersects with Passengers but not the Autonomous circle.

An important consideration is the situation where the remote pilot loses its command and control (C2) link to the RPAS. This "link loss" situation means the vehicle is entirely on its own in terms of its future decision making, until the link is restored. A spectrum of link loss behaviours could be envisaged depending on the sophistication of the UAV.

Very basic behaviour could be "hover in position" until link is restored and descend vertically on low power, and its left to provenance as to what harm befalls other air users unexpectedly encountering the vehicle or those underneath the UAV at the time of its descent.

At the other end of the spectrum, may be an automated "return to base" or "navigate to safe zone" response, where the vehicle makes decisions on navigating around obstacles, avoiding other air users etc. to reach a predetermined location. This would be classed as fully automatic or even autonomous if, for example, the vehicle makes decisions on the safest place to land, perhaps using machine learning techniques to differentiate between, say, an empty field and a busy road.

For this reason, it is difficult to constrain any BVLOS UAV discussion purely in terms of RPAS. This paper will therefore use the terms UAV and UAS and of course drone depending on the level of formality required.

2 The Promise of BVLOS

Being able to fly a UAV Beyond Visual Line Of Sight (BVLOS) opens up a vast array of new use cases and provides a means to greatly improve the efficiencies of existing operations that either require crewed vehicles such as helicopters, or are highly labour intensive such as linear inspections of infrastructure[2]. The market for BLVOS operations is expected to usher in (PWC 2022) a new multi-billion pound industry globally, and the types of operations that are being envisaged will radically change the nature of our skies. One of the most conspicuous usages will be package delivery, where drones are used to deliver small parcels that are currently distributed by road vehicle. Amazon (The Verge 2022) for example, are actively progressing this with plans to trial the technology in the Californian town of Lockeford.

Fig. 2. Amazon Package Delivery Prototype

---

[2] Linear inspections can be done currently by a VLOS UAS but there are practical difficulties in ensuring the UAV is kept within 500m of sight. Not only does the ground crew continually have to move after every segment, but there may be access difficulties along the way (e.g. private property, inaccessible terrain etc). Extended Visual Line of Sight (EVLOS) is also permitted by regulators where other ground crew can maintain line of sight other than the pilot. This helps to some extent but would still be impracticable for very long inspections.

Japan's national postal services is also aiming to start postal deliveries in April 2023 based on a new type of delivery drone.

Fig. 3. Mail Delivery Drone in Japan

While the delivery of goods directly to each household in congested urban areas may present some challenges given the unpredictable nature of the landing area the drone may encounter, the ability to deliver to a local distribution centre such as a supermarket or postal office seems a readily achievable first step.

Another disruptive technology that BVLOS operations will enable is Urban Air Mobility (UAM) that will offer affordable uncrewed airborne taxis to passengers in urban areas at a cost per mile commensurate with road vehicles. Boeing has laid out a Concept of Operations for UAM (Boeing 2022) and the model is not dissimilar to the vision of other air carriers such as Embraer in their Urban Air Traffic Management Concept of Operations (Embraer 2020). Initially, UAM will be realised through piloted vehicles, but these will need to align with decarbonisation objectives for the much larger volume of predicted air traffic operating in an urban environment. There is therefore active work to deliver the next generation of electric air vehicles that will be required to support UAM. For example, in the UK, Vertical Aerospace Group Ltd is developing an electric Vertical Take Off and Landing (VTOL) aircraft called the VX4 (Vertical 2023). There is therefore real credible and tangible work being undertaken and EASA[3] has produced a VTOL Special Condition and Means of Compliance (EASA 2022) to support the development of this novel type of vehicle. Although this will be piloted initially, these vehicles open the door to uncrewed UAM in the future.

---

[3] The CAA are also using this as a certification basis although the FAA is taking a slightly different approach (SUAS News 2022).

Fig. 4. Embraer Vision for a Vertiport

Similar to a taxi rank, passengers will travel by conventional means to a "Vertiport" and then take a UAV to another vertiport located in another part of the city. The UAM vehicles will need to integrate with other crewed air users such as helicopters as well as other smaller UAVs potentially operating in larger volumes, and this will complicate the role of air traffic management. The burden of air traffic control is likely to be eased through strategic as well as tactical mitigation; *strategic* in the sense that UAVs may well travel along predefined air corridors that crewed aircraft will avoid, and *tactical*: through dynamic air traffic management involving the monitoring and actively maintaining separation of all aircraft.

The rich future for BVLOS is really the "Holy Grail" for UAV operations so why isn't everybody doing it now? The simple answer is: *the commercial viability of meeting regulation.*

3 BVLOS Regulations

3.1 UAV Regulatory Categories

In 2015 EASA (EASA 2015) proposed the concept of drone categories to help in the regulation of UAS. This was seen as necessary as the traditional methods for certifying manned aircraft cannot be applied in the same way for remotely piloted aircraft. For example, for manned transport the focus is on preserving the safety of those on board and less emphasis is given to uninvolved parties on the ground. There is also less emphasis on the operational context. For drones, there is a complete shift of emphasis with the risk being entirely to non-involved parties both on the ground and in the air. Ground based risk assessments are highly dependent on the operational context, for example, risks are different if the drone is flying over the sea as opposed to a city and so operating context was seen as being a key consideration.

The EASA approach, which has been adopted by the CAA (CAA 2022), is to classify drones into 3 categories:

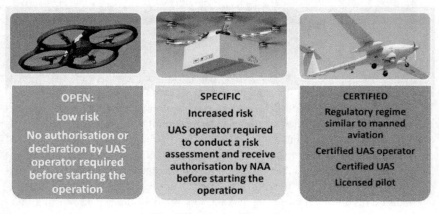

Fig. 5. Regulatory Categories

Open (Low Risk): The 'open' category operation is low-risk and for simple drone operations, where the risk to third parties on the ground and to other airspace users is mitigated through operational limitations. The 'open' category drones do not require an authorisation by a National Aviation Authority (NAA) for the flight, but should stay within defined limitations for the operation (e.g. safe distance from aerodromes, from uninvolved persons). This category of operations is only subject to a minimal aviation regulatory system, focussing mainly on defining the limits of such a category of operations. No certification, approval, licence or other equivalent document is required in relation to the operation of drones, except in the case of more complex, low-risk operations where adequate knowledge and skills need to be demonstrated.

Specific (Medium Risk): When an operation poses more significant aviation risks to persons overflown or involves sharing the airspace with crewed aviation, the operation is placed in the 'specific' category. The 'specific' category requires an Operation Authorisation (OA) issued by an NAA with specific limitations adapted to the risk posed by the operation. For these activities, each specific aviation risk is analysed and adequate mitigation means need to be agreed by the NAA before the operation can start, based on a safety risk assessment. The approval is materialised with the issue of an OA to the operator.

Certified (High Risk): Certification is required for operations with an associated higher risk due to the kind of operation, or might be requested on a voluntary basis by organisations providing services (such as remote piloting) or equipment (such as detect and avoid). When uncrewed aviation risks rise to a level similar to normal, crewed aviation, the operation is placed in the 'certified' category of operations.

These operations and the drones involved therein are treated in the classic aviation manner: multiple certificates are issued (as for crewed aviation) plus some certificates specific to drones. The operations in the 'certified' category are for drone operations with a high risk and with a wider scope of operation than the 'specific' category.

Examples are international cargo transport operations with large UAVs, transport of persons or any other operation where the risk assessment process of the 'specific' category does not sufficiently address the high risks involved in the operation.

4 Current State of UAV Operations

4.1 Open Category (Low Risk)

The mainstay of the risk management approach for open category operations is to retain visual line of sight of the vehicle and to operate sufficiently far away to avoid harm to people if something goes wrong. At a high-level, the approach is:
- The bigger the drone that is being flown (up to a maximum of 25kg), the further the operation needs to be away from "uninvolved" people (e.g. members of the public as opposed to the ground crew);

- Limiting the maximum height above ground to 120m;
- Flying a maximum of 500m from visual line of sight of the pilot or a ground crew observer.

For many business cases, this is perfectly adequate – for example, if using a UAV to inspect a local asset such as a building or aerial filming in sparsely populated locations. Regulators also provide predefined risk assessments for some of these types of operations to help smooth the route to achieving operational approval. As of April 2022, in the US alone (Statica 2022), there were over 300,000 registered commercial drones, so even with these restrictions, there seems a significant market for VLOS operations.

The limitation of maintaining VLOS to 500m is nevertheless a significant constraint that, at a practical level, rules out an immense number of use cases where BVLOS is essential such as remote package delivery, search and rescue operations, linear inspections and geographic surveys.

4.2 Certified Category (High Risk)

BVLOS operations fall in either the specific or certified categories of operation as, by definition, visual line of sight cannot be maintained. As the certified category provides the operator with the wider ability to fly the aircraft much in the same manner as crewed aircraft, insight into how widespread adoption of certified BVLOS operations can be gained from a view of those aircraft providing position reporting information.

The figure below provides a brief snapshot on a Saturday afternoon in December 2022 of those aircraft transmitting position reports and reported on flightradar24.com.

Fig. 6. UAVs Transmitting Position Reports 10th Dec 2022.

As can be seen, from almost 12,000 aircraft transmitting position reports globally, only 2 were registered as drones from which it's reasonable to infer very low levels of certified drone use. The reason for this is that the regulatory requirements for certified operations require the vehicle to be assured to the equivalent standard as crewed aviation, which sets the bar very high for current drone operators that tend to be Small and Medium-sized Enterprises (SME). For example, standards for certified operations that are emerging such as DO-377A (RTCA 2019), which defines the minimum performance requirements for the C2 Link system, needs to consider a wide range of scenarios such as taxiing, take-off from busy airports and flights in controlled airspace, integrating with crewed aviation so there will be hazards frequently with catastrophic outcomes that will require the highest levels of development assurance rigour.

The CONOPS vision from Boeing and Embraer acknowledge that a significant amount of additional infrastructure will be required for these types of operation to be fully realised and only low densities of aircraft are considered currently possible with piloted UAM vehicles using conventional ATM procedures and technologies. Possibly in this decade, we may see medium-density UAM operations but only with the support of new ATM procedures and technologies to meet the demand for operations. The expectation is that it will be 2030 and beyond before high density operations will be routinely supported.

4.3 Specific Category (Medium Risk)

At a more practicable level, there is the specific category of operation, which has been the focus of regulatory and standards development in recent years. The Joint Authorities for Rulemaking of Unmanned Systems (JARUS) is a group of experts from the National Aviation Authorities (NAAs) and regional aviation safety organisations. Its purpose is to recommend a single set of technical, safety and operational requirements for the certification and safe integration of Unmanned Aircraft Systems (UAS) into airspace and at aerodromes.

JARUS worked on an acceptable means of compliance for those seeking approval for operations in the specific category. A key output of their work was the Specific Operations Risk Assessment (SORA) (JARUS 2019) that provides a methodology for demonstrating to an approval authority that the system is sufficiently safe for its intended use.

Specific category approvals of BVLOS operations, while by no means commonplace, are proving more readily achievable. Specific category approvals require the exact operational use case to be defined and for the ground and air risks to be explicitly identified, assessed and mitigated. The disadvantage is of course that operational approval will only be for the use case defined, and so it is not a license to fly the UAV anywhere in any circumstance, but because the operation is well-defined, the advantage is that the air and ground risks can be more readily assessed and mitigated to the satisfaction of the NAA.

An example of this type of operation has been the use of the Windracers UAV for Royal Mail deliveries to remote islands. The initial routes identified for the new service include the Shetland Islands, Orkneys, the Hebrides and the Isles of Scilly. Flights have begun with a service trial from Lerwick to Unst in the Shetland Islands. Over the next three years, the aim is to secure more than 50 further routes supported by up to 200 drones. Longer term, the ambition is a fleet of more than 500 drones servicing all corners of the UK and beyond.

Fig. 7. Royal Mail deliveries to Scottish Islands

Here, the ground risks are clearly mitigated by operation in sparsely populated areas, and, as typical for operations in the specific category, a well-defined "operational volume" is defined in which the UAV will be contained or *segregated*. These form strategic mitigations for ground and air risks that help support the safety case for the operation.

The use of segregation can also be employed effectively in some circumstances and companies like the National Grid and Network Rail are exploiting this as part of their infrastructure inspection programmes. The National Grid (NG) owns approximately 22,000 steel lattice pylons that carry overhead transmission conductor wires over 7,200km in England and Wales. Transmission pylon steelwork condition can deteriorate through corrosion, so periodic assessments are made to understand the health of the network (National Grid 2022). NG inspects around 3,650 steel lattice pylons each year, capturing high definition still colour images of steelwork and is now trialling a CAA approved BVLOS operation in conjunction with the operator sees.ai (Sees.ai 2022), not only enabling automatic flight operations from a central pilot control station, but also, using AI/ML techniques to analysis the resulting data to rapidly increase the speed at which issues can be detected. BVLOS operations are more advanced with NG and others like National Rail (National Rail 2022), as the transmission network effectively forms a segregated airspace. Other airspace users are already segregated from these areas, the land is owned by these organisations and in the case of NG, the grids are usually over sparsely populated areas.

The principle of segregation can be taken forward to create virtual highways for drone operations. Altitude Angel is one of the first UAS Traffic Management (UTM) companies (also referred to as a U-Space Service Provider in European territories) providing early UTM services with their own segregated BVLOS trial area – the Arrow Drone Zone (Drone Safety Map 2022) as highlighted in the shaded horizontal area below (the circular areas on the far left and right are the landing/takeoff areas).

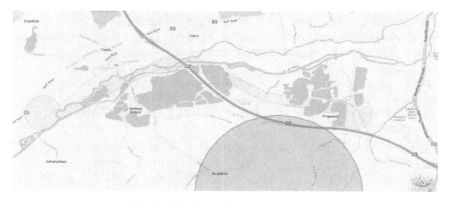

Fig. 8. Altitude Angel Arrow Drone Zone

Altitude Angel is leading a consortium of businesses to extend this concept to build and develop 165 miles (265km) of 'drone superhighways' connecting airspace above Reading, Oxford, Milton Keynes, Cambridge, Coventry and Rugby over the next two years (Altitude Angel 2022). Segregation does of course have its limitation; not everywhere will be suitable for segregation and it won't be suitable where the geography for BVLOS operations cannot be dictated, and so is not a complete solution for large scale adoption of widespread BVLOS operations at a national level.

5 The Goldilocks Syndrome

We find ourselves in a Goldilocks situation. The open category provides a relatively straightforward path to commercial operations but is too constraining for the types of BVLOS operations that will fully ignite growth in the UAV industry. Fully certified operations on the other hand offer much in terms of the desired capability, but is many years away from being commercially achievable by most, and having the required level of technical ATM support to realise high volume operations.

Even in the middle ground of specific category, operations have been approved on the basis of significant strategic mitigations of ensuring operations are (mostly) remote and far from people. These developments are of course positive and set a precedent for similar types of operations, but still, in a sense, are skirting around the problem; what if your specific use case cannot be operated over open seas or rural countryside? Indeed, what if we do have a specific category operation that is near populated areas and even near airports and controlled airspace?

The SORA process, as previously introduced for assessing risk in the specific category, categorises operations into 6 levels of risk called the Specific Assurance and Integrity Level (SAIL), which are expressed as roman numerals I, II, III, IV, V and VI from lowest risk to highest. The following table shows how the assessment of ground risk and air risk is combined to provide an overall SAIL value, which then dictates the integrity and assurance requirements that need to be met for the level and importantly, the level of rigour that needs to be applied in meeting those requirements (called "robustness" in SORA). A SAIL of I and II can be considered low risk, III and IV, medium risk and V and VI high risk.

Table 1. How the assessment of ground risk and air risk is combined to provide a SAIL value

Ground risk value	Residual air risk value			
	ARC-a	ARC-b	ARC-c	ARC-d
<=2	I	II	IV	VI
3	II	II	IV	VI
4	III	III	IV	VI
5	IV	IV	IV	VI
6	V	V	V	VI
7	VI	VI	VI	VI
>7	SORA process not suitable			

Again, we have another Goldilocks situation with low-risk SAIL values allowing BVLOS operations but likely only when operations are away from uninvolved persons. High risk SAIL operations would offer a far greater range of operations, but the level of assurance required is verging on the same required for certified operations and so is not likely to be practicably achievable in the short term, at least for SME operators.

The medium risk SAIL operations (SAIL III and IV) do however offer a promising opportunity for conducting operations that are sufficiently diverse to meet many commercial needs, yet have enough commercially achievable strategic and tactical mitigations to manage the risk. The path is by no means an easy one, but it is one that is in reach if we are prepared to grasp it.

6 Medium Risk Specific Category Operations

Consider the following hypothetical scenario to illustrate the types of strategic and tactical mitigations required for routine BVLOS operations. There are a number of large product distribution centres to the west of Bristol, UK, which are involved in the fulfilment of online orders for goods typically less than 5kg in weight. Rather than shipping these small goods via road vehicle, a company is considering using drones to deliver to more local distribution centres in the Bristol area, such as supermarkets. Although mainly rural, the routes will not be segregated, will travel over busy roads and motorways and near urban areas with significant presence of groups of people such as shoppers. The area is also near Bristol Airport and so will be operating through Class D airspace of the airport's Control Zone (CTR).

The SORA[4] assessment determines that from the resulting ground and air risks, the SAIL for this operation is SAIL IV. This is the high end of the medium risk scale as there is no segregation – the drone may well fly over assemblies of people and although operations will be flown below 400ft, the risk of a drone "flyaway" (uncommanded flight to possibly higher altitudes) could bring the vehicle into conflict with a crewed aircraft near the airport. Clearly, some of these hazards will be "catastrophic" and so what kind of UAS would be appropriate for regulatory approval of this type of operation?

While a complete means of compliance is still being developed, the general shape of the UAS characteristics that are considered necessary are starting to emerge. Firstly, consider what characteristics are not considered necessary. For the medium risk specific category, having type certification for each of the components in the UAS, such as the vehicle, link and control station are not essential; these are in the realms of the certified category. While having a type-certified vehicle, for example, may help support a specific category safety case it is not mandatory; rather, it is the safety of the UAS as a whole that is important and the associated safety justification for the system rather than its individual components is key. The general UAS regulatory requirements coupled with specific requirements that come from the SORA process give a good indication of the types of technology that will be required. These fall broadly into the categories:

- Route planning and environment awareness;
- Command and Control (C2) Links;
- UAS Traffic Management;
- Positioning, Navigation and Timing (PNT);
- Detect and Avoid.

---

[4] This is based on SORA version 2.0. It should be noted that version 2.5 of the SORA guidance was issued for consultation in Dec 2022 and is currently open to industry review until 6th March 2023. This particular assessment is therefore subject to change but for the purposes of this paper, the consideration is for operations in the SAIL IV category.

The following sections describe these in more detail and provides an assessment of the relative readiness of these technologies.

6.1 Route Planning and Environmental Awareness

For a number of reasons, BVLOS operations will not be flown by pilots in the same way that conventional aircraft are flown. Although it is possible for a pilot to have a first-person view from the aircraft, the CAA do not permit this as a valid method of safely piloting the vehicle and when geostationary SATCOM-based C2 links are used, it becomes impracticable for a pilot to fly the vehicle using direct control of the aerodynamic surfaces due to link latency. Rather, the aircraft itself will conduct missions in an automatic manner with the pilot first plotting the route via waypoints on a map and the UAV's flight management system automatically following the route using information such as its current position, velocity and direction and making all the required adjustments to keep on course, despite possible prevailing wind conditions. As the pilot will effectively be flying under Instrument Flight Rules (IFR), significant route planning will be required, including determination of the route geography, that is, establishing a profile of the physical terrain that may present obstacles and hazards both permanently (like hills and buildings), and temporary, such as cranes or areas that have been virtually "fenced" due to other air user activities.

Many UAV manufacturers already have vehicle and ground control systems that implement this form of automatic route planning and execution. The figure below shows a typical user interface (reproduced with permission from SPH Engineering 2022):

Fig. 9. Example Route Planning Software – UgCS

These applications will be developed to commercial standards, perhaps from companies with ISO9001 quality management certification. The SORA process for SAIL IV operations requires software to be developed to a standard agreed as appropriate with the NAA. What is appropriate will be dependent on the specific use case and how errors in the planning system may give rise to hazards, but it seems likely that development to a standards-based assurance level in accordance with ED-109A (ground) (EUROCAE 2012) and DO-178C (air) (RTCA 2011) would help support the safety case[5].

6.2 C2 Links

Fundamental to managing the risks of a UAV is being able to communicate and instruct the vehicle to take particular actions, whether this is to simply start a mission, take an alternative route or course of action when instructed to do so, say to respond to an air traffic control instruction to deconflict a possible encounter with another vehicle, or even to issue a "kill" command to terminate the mission in the event of a "flyaway". A C2 Link could take the form of a highly reliable single link or, what is more likely, a multilink configuration using diverse technologies to help mitigate the risk of common mode failures. Such a multilink design not only helps with safety risks but also helps mitigate security risks so if one link technology is compromised, control can be passed to another. A multilink configuration also means that a high degree of reliability can be formed from a set of links that have lower reliability, a configuration that might be more practicable to implement using publicly available link technologies – thus making the UAS a more commercially viable proposition. One such configuration is to have C2 Link formed of a SATCOM-based link and a terrestrial mobile network from a commercial Mobile Network Operator (MNO), providing LTE (4G/5G) based radio communications. This diversity not only provides redundancy but also flexibility in coverage. For example, at low altitudes in urban areas, the buildings lining streets can form a "canyon" that can interfere with SATCOM reception with reflections giving rise to "multipath" that can degrade the signal quality. In these cases, the terrestrial mobile network is likely to be more reliable. In more rural locations where terrestrial mobile reception may be patchier, the SATCOM link will be preferable.

Note however, that as the UAV is following a predetermined route automatically, the C2 uplink is not as critical as one might initially imagine. It is in fact the C2 downlink from UAV to the pilot that is more critical, as this is where information about the UAV's position, direction, altitude and status is reported to the pilot. For example, if the UAV is straying from position or some physical

---

[5] ED-109A equates to DO-278A and ED-12C equates to DO-178C.

malfunction has occurred, then the pilot will need to know this quickly in order to take appropriate action to maintain safety.

While it is technically currently possible to use SATCOM and LTE communications[6] for C2 links, changes will need to be made to support UAVs in volume. LTE communications pose particular challenges as given the improved line of sight an airborne vehicle has over a ground-based mobile phone, a UAV can connect to multiple towers and estimates show that even a moderate number of airborne vehicles will cause interference with normal land-based communications (OFCOM 2022). Modifications to the 3GPP specification have been put in place to better support UAVs by terrestrial mobile network operators and there is already an LTE "aerial profile" (AJCA 2020) and one for 5G will be developed in the near future. These define the subset of the 3GPP specifications considered essential to launch interoperable services.

A number of service products are also considered necessary from SATCOM and MNO operators that move these communication providers from simple generic datalink providers to C2 Communication Service Provider (C2CSP). These might, for example, include awareness that a particular channel is carrying C2 data for the purposes of prioritisation, position reporting of vehicles and provision of service status information such as planned and unplanned coverage forecasts.

6.3 UAS Traffic Management (UTM)

As the UAV is being flown under IFR, the pilot will be unaware of the plans that other air users have that may conflict with their intended vehicle's route. While the UAV may have tactical mitigations to avoid collisions (e.g. Detect and Avoid), this should only be used by exception, and more strategic separation techniques should be adopted. These mitigations take the form of a UTM provider (also known as a U-Space Service Provider – USSP), which will be one of many companies offering traffic management services specifically for UAVs. Current air traffic management primary radar is not able to detect small drones and so a service geared towards UAVs is required. While crewed aircraft traffic is managed by a single Air Navigation Service Provider (ANSP), this will not be the case in the UAS world where multiple UTMs will operate with a single Common Information Service Provider (CISP) acting as an information backbone for sharing dynamic traffic information in real time. Each UAV operator will submit flight plans to a UTM provider and the UTM, using CISP supporting services, will approve or deny the flight. As well as these strategic planning mitigations,

---

[6] As of 2022 it was not permitted to use MNO communications for UAV BVLOS operations but this is being relaxed through the introduction of spectrum licensing by OFCOM (OFCOM 2022), which will allow operators to apply from January 2023.

the UTM provider will also implement tactical mitigations mid-flight. For example, if a non-cooperative UAV is detected, an operator can be instructed to alter their vehicle's current route to remain well clear.

UTM services might, for example, provide flight plan approvals, tactical air traffic management as well as other information such as terrain, weather and geofencing information for route planning. Unless within an existing airports Aerodrome Traffic Zone (ATZ) where the air traffic controller already has obligations to ensure separation (c.f. Cranfield 2021), the UTM services will be provided on an information only basis. In the future however, there will be more reliance on these services for flight safety, and so it is expected that UTM providers will need to become certified in some way and hence all critical hardware and software applications approved for use. The other entity, the Common Information Service Provider (CISP) previously mentioned, is a relatively new concept but will likely take the role of sharing information between UTMs and ANSPs at a national level. UTM is still in its infancy but there are many demonstration locations globally providing test facilities (EUROCONTROL 2021) and the UK are already looking at how to resolve UTM aspects with the CAA, NATS and industry experts. There are also proposals for low level tracking of drones around aerodromes in discussion.

6.4 Position, Navigation and Timing (PNT)

A UAV will only follow its planned route correctly if it has accurate information on its current position, velocity and direction. As routes are not segregated in the example, there is no "safety buffer" in which the UAV may stray and so it's essential that a PNT solution, such as one commonly implemented using a Global Navigation Satellite System (GNSS) is sufficiently reliable. For use cases such as package delivery, there may be more flexibility around position accuracy during cruise phases. Standard GNSS services (GPS, Galileo etc) may be sufficient in these cases when accompanied by an augmentation service such as EGNOS (EUSPA 2022) to ensure safety-of-life accuracy of around 1m.

However, use cases that involve surveys and mapping (e.g. orthophotography and photogrammetry) may require centimetre accuracy and so additional positioning technology will be required such as integrated Inertial Measurement Units (IMU).

6.5 Detect and Avoid (DAA)

One of the fundamental requirements arising from regulation is that UAVs should meet the "rules of the air"[7], which includes, amongst other requirements the ability to avoid other air users. For UAVs with no human pilot on board, the ability to "see and avoid" or remain well clear of other air users can only be met through technological means[8]. While UTM providers will be able to deconflict other air users that are registered with the service or transmitting their presence through conspicuity technologies such as FLARM (FLARM 2022) and ADS-B[9] (Wikipedia 2022a), there are still some air users that may not have these technologies such as paragliders[10] and balloons[11], and hence the ability for the UAV itself to detect and avoid other air users, whether cooperative or not, seems a likely tactical mitigation.

DAA takes a number of forms. On the most fundamental level, DAA will prevent basic collision avoidance and many consumer vehicles will already have this capability; for example, if the UAV is inadvertently commanded to fly into the ground or another physical obstacle like a wall or building, then the vehicle will automatically take action to prevent the collision.

The next level of DAA operation is to anticipate future conflicts with other air users who may be converging on paths that will bring both vehicles into conflict. The DAA system might initially inform the pilot of the potential conflict and in more sophisticated functional implementations, automatically take avoidance action independently of the pilot instruction. Note that this technology is not some future aspiration but there are examples of technical implementations now, for example, the Honeywell's IntuVue RDR-84K system underwent trials in 2022 (Honeywell 2022). What is missing however, is the regulatory approval – at the time of writing, the CAA has not approved any DAA subsystem[12].

---

[7] Article 7 point 2 of the CAA's Implementing Rule CAP1789A (Dec 2022) requires compliance to Commission Implementing Regulation (EU) No 923/2012.

[8] As mentioned earlier, the CAA does not consider the video transmission of a first-person pilot view to be sufficient to meet this requirement.

[9] Note that standard ADS-B transmissions are not permitted below 500ft due to potential risks of spectrum congestion so other frequencies or other technologies will be required to address these issues.

[10] Paragliders are increasingly adopting FLARM. One of the problems is the proliferation of incompatible electronic conspicuity standards: ADS-B, FLARM, Pilot-Aware, etc.

[11] And of course, other natural air users such as flocks of birds.

[12] The CAA has published a standard for uncertified ADS-B: CAP1391: Electronic conspicuity devices (CAA 2021) such as the uAvionix's SkyEcho (UAVIONIX 2023).

7 So How Close Are We?

From the previous roundup of the technologies required to support medium risk operations, much of the technology that is required is already available although perhaps more additional rigour will be required in assuring the hardware and software systems to better support a safety case. This might, for example, involve assuring software to an aviation standard rather than commercial quality standards. Other areas will require further trials and the development of the evidential basis that will provide a regulator sufficient confidence in the overall safety of the system. For example, UTM and DAA technologies are novel in this context and so will require careful argumentation to convince the regulator of their reliability. However, none of these seem technically insurmountable or particularly far away. One problem, however, is that there are many activities underway from a wide range of stakeholders that will each eventually be seeking approval from a regulator with limited resources.

There is a real paradigm shift with the regulatory landscape for UAS; the industry is moving from a handful of large airframe manufacturers seeking approval of aircraft over time periods measured in years, to potentially hundreds of smaller organisations seeking approval for their own unique set of circumstances, which can be vastly varied and diverse, both in the specific type of operation and the set of equipment that form the UAS.

8 Conclusion

UAS operations broadly fit into three categories:
- the low risk "open category" operations where line of sight is maintained and operations are well away from people;
- the certified use where operators are certified to fly any vehicle anywhere as the vehicle and all dependent subsystems have been certified through independent verification;
- the specific category; where a specific use case scopes the operation and a safety case is developed for the system as a whole for that particular use alone.

A Goldilocks situation is emerging with the low-risk open category offering some utility but is ultimately limiting a wider range of use cases by not being appropriate for BVLOS operations. On the other side of the spectrum, certified operations will require a large amount of infrastructure to support and will be costly putting the development assurance activities beyond most of the commercial organisations wishing to move into the market.

The middle ground of the "specific" category seems to be "just right" and there has been some quick wins in terms of specific operations using BVLOS in rural and segregated airspace. The specific category however offers a much richer vein of opportunity and the technological support required to provide safe operations for medium risk operations seem mostly available now with the remainder within grasp in the next few years.

Acknowledgments The author would like to thank Dewi Daniels and Ross Hannan for their valuable review comments.

Disclaimers All views expressed in this paper are those of the author.

References

AJCA(2020) LTE Aerial Profile version 1.00, https://www.gsma.com/iot/wp-content/uploads/2020/11/ACJA-WT3-LTE-Aerial-Profile_v1.00-2.pdf, accessed Dec 2022

Altitude Angel (2022) https://www.altitudeangel.com/news/uk-consortium-reveal-blueprint-to-build-165-mile-drone-superhighway, accessed Dec 2022

Boeing (2022) Concept of Operations for Uncrewed Urban Air Mobility, https://wisk.aero/wp-content/uploads/2022/09/Concept-of-Operations-for-Uncrewed-Urban-Air-Mobility.pdf, accessed, Dec 2022

CAA (2021) CAP 1391: Electronic conspicuity devices, 25th Feb 2021

CAA (2022) CAP 722 Unmanned Aircraft System Operations in UK Airspace – Guidance, v9.1, 22 Dec 2022

Cranfield (2021) https://www.cranfield.ac.uk/press/news-2021/cranfield-becomes-first-airport-to-deploy-altitude-angels-guardian-enterprise-platform, accessed Dec 2022

Drone Safety Map (2022) https://dronesafetymap.com/, accessed Dec 2022

EASA (2015) A-NPA 2015-10, Introduction of a regulatory framework for the operation of drones, 31 July 2015.

EASA (2022) Special Condition for VTOL and Means of Compliance, https://www.easa.europa.eu/en/document-library/product-certification-consultations/special-condition-vtol, accessed Jan 2023

Embraer (2020) Urban Air Traffic Management Concept of Operations

EUROCAE (2012) Software Integrity Assurance Considerations For Communication, Navigation, Surveillance And Air Traffic Management (CNS/ATM) Systems

EUROCONTROL (2021) European network of U-space stakeholders, https://www.eurocontrol.int/product/european-network-u-space-demonstrators, accessed Dec 2022

EUSPA (2022) https://www.euspa.europa.eu/european-space/egnos/what-egnos, accessed Dec 2022

FLARM (2022) https://www.flarm.com/, accessed Dec 2022

ICAO (2015) Doc 10019: Manual on Remotely Piloted Aircraft Systems (RPAS)

Hampton, Pugh and Ball (2020), "Developments in Safety & Security Integration: Remotely Piloted Unmanned Aircraft Systems Command and Control". Assuring Safe Autonomy: Proceedings of the 28th Safety-Critical Systems Symposium (SSS'20) York, UK, 11th-13th February 2020

Honeywell (2022) Autonomous Detect-and-Avoid Demonstration, https://www.youtube.com/watch?v=DBgq8qivRdU, accessed Dec 2022

JARUS (2019) guidelines on Specific Operations Risk Assessment (SORA), JAR-DEL-WG6-D.04, v2.0

National Grid (2022) https://www.nationalgrid.com/age-ai-national-grid-trial-futuristic-automated-corrosion-inspection-electricity-transmission, accessed Dec 2022

National Rail (2022) https://www.railway-technology.com/news/network-rail-bvlos-drone-flight, accessed Dec 2022

OFCOM (2022) Spectrum for Unmanned Aircraft Systems (UAS), Approach to authorising the use of radio equipment on UAS, Consultation document June 2022

PWC (2022) https://www.pwc.co.uk/issues/emerging-technologies/drones/the-impact-of-drones-on-the-uk-economy.html, accessed Dec 2022

See.ai (2022) https://www.sees.ai/about-us/, accessed Dec 2022

Statics (2022) https://www.statista.com/statistics/1221517/uas-drone-registrations-united-states/, accessed Dec 2022

Royal Mail (2022) https://www.royalmail.com/sustainability/environment/drones-connecting-remote-communities-across-the-uk, accessed Dec 2022

RTCA (2011) DO-178C, Software Considerations in Airborne Systems and Equipment Certification

RTCA (2019) DO-377A – Minimum Aviation System Performance Standards for C2 Link Systems Supporting Operations of Unmanned Aircraft Systems in U.S. Airspace

SAE (2021) Taxonomy and Definitions for Terms Related to Driving Automation Systems for On-Road Motor Vehicles J3016_202104, accessed Dec 2022

SPH Engineering (2022), UgCS flight planning software by SPH Engineering (http://www.ugcs.com)

SUAS News (2022) https://www.suasnews.com/2022/12/faa-nprm-joby-g-1-issue-paper-docket-faa-2021-0638-caa-comments, accessed Jan 2023

The Verge (2022) https://www.theverge.com/2022/6/13/23165727/amazon-drone-delivery-pilot-lockeford-faa, accessed Dec 2022

UAS Vision (2022) https://www.uasvision.com/2022/12/09/japan-post-to-start-drone-deliveries/, accessed Dec 2022

UAVIONI (2023) SkyEcho, Electronic Conspicuity, https://uavionix.com/products/skyecho, accessed Jan 2023.

University of Southampton (2020) https://www.southampton.ac.uk/news/2020/05/drone-trial-delivery.page, access Dec 2020.

Vertical (2023) Vertical Aerospace Group Ltd, VX4 VTOL aircraft, https://vertical-aerospace.com/vx4, accessed Jan 2023

Washington Times (2022) https://www.washingtontimes.com/news/2022/nov/30/drones-will-now-be-uncrewed-not-unmanned-pentagon-/, accessed Dec 2022

Wikipedia (2022) "Unmanned Aerial Vehicle", https://en.wikipedia.org/wiki/Unmanned_aerial_vehicle, accessed Dec 2022

Wikipedia (2022a) Automatic Dependent Surveillance – Broadcast, https://en.wikipedia.org/wiki/Automatic_Dependent_Surveillance%E2%80%93Broadcast, accessed Dec 2022

How Nuclear New Builds Incorporate Lessons Learned

Tom Hughes

EDF Energy

Abstract *Ask a member of the public to think about 'Nuclear Safety' and it's highly likely a number of emotive words and images will spring to mind – events like Three Mile Island, Chernobyl and Fukushima have touched thousands, tainted the industry's reputation and contributed to years of slow progression of the technology. With nuclear power offering a viable solution to many of the planet's most pressing concerns such as climate change, fossil fuel dependence and energy security, it's time to change the narrative a little and think about how the industry has addressed the challenges raised by these significant events. In this presentation, Tom will provide an insight into how the nuclear industry rigorously assesses and implements changes based on operational experience, not only to physical designs but also to the way plants are organised and operated. Several fascinating case studies will be explored and explained from a technical perspective, as well as looking at how human factors influence safety.*

© Tom Hughes 2023.
Published by the Safety-Critical Systems Club. All Rights Reserved.

Can software engineering methods give us better software safety standards?

James Inge

Defence Equipment & Support

Bristol, UK

Abstract *Are safety assurance standards actually software engineering artefacts, part of the decomposition of organisational goals into software requirements and designs? Loosely speaking, aren't they just software that is executed by an organisation rather than a computer? And if so, can we use software engineering methods to improve them? Software safety standards have a vital role in delivering safe products, services and systems. In critical systems, software failures can lead to significant loss of life, so it is especially important that such standards are well understood by their users. Yet, they are often verbose, lengthy documents written by committees; hard for the uninitiated to immediately digest and understand, and awkward to implement as written. This implies that the review process for such standards is not entirely effective. Building on the author's MSc research at the University of Oxford, this paper examines how techniques from the domain of software engineering and allied fields can be used to improve the review of standards, potentially leading to better safety standards and safer systems. It presents a selection of potential techniques, evaluates the results of applying them to Def Stan 00-055, (the Ministry of Defence's Requirements for Safety of Programmable Elements in Defence Systems), shows how they can be helpful, and discusses the practicalities of applying them to review of new and existing standards.*

1 Introduction

1.1 Why standards for software safety assurance are important

With software playing an ever-increasing role in delivering the functionality of critical systems, evidently it is important to ensure that software will operate safely. For critical systems, it is also important to gain confidence that this will

© James Inge 2023.
Published by the Safety-Critical Systems Club. All Rights Reserved.

be the case before deploying the system. Failing to adequately plan to achieve this can be a notable cause of cost rises and delays to major projects.

A well-documented example of such delays was the UK Ministry of Defence (MOD) procurement of the Chinook Mk 3 helicopter (Fig. 1). Eight Mk 3s were delivered to specification by Boeing in 2001 at a cost of some £259 million. As the avionics software for their bespoke digital 'glass cockpit' could not be certified to meet UK military airworthiness standards, they could not be used in operations until 2009 (Burr 2008). The problem was not that the software was known to be unsafe, but that it was not known to be safe. The MOD could not demonstrate its safety as it had not contracted for Boeing to provide either sufficient evidence of safety analysis, or access to source code that the MOD could analyse itself (Bourn 2004). The issue was resolved by first reverting the Mk 3 Chinooks to an earlier, proven design standard at a cost of over £90 million, then later upgrading them to a different type of glass cockpit. The project for this upgrade was itself delayed a further nine months due to software development issues (Morse 2013).

Fig. 1. Chinook Mk 3 (© Crown copyright 2016)

This Chinook example demonstrates that it is important for software not just to work, or even just to work safely: it needs to be demonstrably safe. Achieving this does not happen by accident. When an organisation needs to acquire new safety-critical software, it needs to communicate its safety requirements and its assurance requirements to the supplier as part of the contract. Using assurance standards for safety assurance such as Def Stan 00-055 (MOD 2021) or IEC 61508 (IEC 2010) is an efficient, repeatable way of doing this that captures accepted good practice, avoids the need for each project or organisation to work up its own requirements from scratch.

1.2 Why quality and review of standards matters

To lead to good outcomes, standards need to have good functional content: what they prescribe needs to be technically effective. As the state of the art develops, standards need maintenance to keep them relevant. It is important to review and update them to incorporate new good practice and remove material which has become outdated. However, other non-functional aspects of standards are also important.

Standards need to be easy for their audience to understand and put into practice, or else risk their technical merit getting lost and the cost of their use rising. If a standard is ambiguous or hard to use, organisations implementing it are likely to budget more to account for the extra time required to understand its requirements and the risk of getting them wrong. If a standard is hard to understand, it is likely to be hard to review and technical problems may go un-noticed. This means that it is desirable for reviews of standards to look at not just their technical merit, but other quality factors that contribute to their practical effectiveness.

1.3 Are software safety standards software (and would it help if they were)?

A case can be made to argue that assurance standards for software safety are in fact software artefacts themselves. Philosophically, they can be seen as sets of instructions, processes and supporting data that are executed by an organisation, rather than a machine (Fig. 2). More practically, by setting assurance requirements for software, they are a part of its high-level specification – part of the decomposition from high-level organisational goals to low-level software requirements.

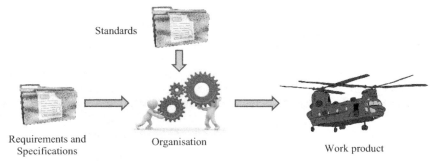

Fig. 2. Standards as an input to the process of an organisation.

According to IEC 61508 (IEC 2010), the definition of 'software' includes 'any associated documentation pertaining to the operation of a data processing system'; and authors such as Ould (1999) and Patton (2005) include specifications among the set of artefacts to consider as part of the software quality assurance and testing processes.

Regardless of whether you accept the argument that standards actually are a type of software, the analogy is helpful. Both software and standards are abstract information products that have important functional and non-functional attributes. It is important for both software and standards to be technically correct to achieve their intent. As documents, it is also important for software code and standards to be easy to understand, so that problems can easily be identified and fixed, and so that they can be maintained efficiently in the future. With these similarities in mind, it is reasonable to ponder whether the discipline of software engineering can teach us lessons for improving standards.

In software engineering, reviews are recognised as an effective way of improving software quality, and a variety of more structured methods are available to help verify and validate development artefacts. In contrast, in the author's experience, formality in reviews of standards and similar documents often extends only to having a process of official committees and meetings that leads to endorsement of a new version.[1] They tend not to be formal in the sense of actually examining the standard in a structured manner, or using formal methods to exploit the structure and semantics of the standard itself as part of the review. Most often, reviewers are simply presented with a draft document and asked to respond with comments. Standards can be lengthy, and reading through and making meaningful comments can be time-consuming for reviewers.

This paper reports on the results of MSc research carried out by the author at the University of Oxford (Inge 2019), investigating whether software engineering techniques could indeed inspire a more effective approach to review of safety assurance standards.

2 In search of a better review method

To attempt to identify a better way of reviewing software safety assurance standards and test the hypothesis that software engineering-inspired methods could provide improvements, the author carried out a literature review to identify potential methods, then applied each of these methods to Def Stan 00-055 Issue 4 (MOD 2016). The results were then compared to evaluate the quantity and type

---

[1] While the ISO/IEC Directives (ISO/IEC 2021) do contain some guidance that can help reviews (see section 2.2.2), in practice, the author has not been aware of this being applied proactively in the committees in which he has taken part.

of issue found by each technique, and the practicalities involved in carrying them out.

2.1 Approaches to evaluating standards

The author's experience of reviewing standards consists mainly of what one might call 'naïve' reviews: reviewers are simply given a text and asked to read through and make comments. An editor or editorial committee determines their resolution and the document is amended accordingly. A meeting or workshop may be held to resolve the comments, or this may be left to the editors. A 'comments form' often guides reviewers to respond in a certain format. The review template used by the British Standards Institute for comments on international standards asks for comments, proposed changes, a reference to the location in the standard's text, and a classification of the comment as general, technical or editorial. The structure of these forms and the presentation of the document under review tends to lead to a particular style of comment. Reviewers read the document sequentially and comment on specific sentences, paragraphs or figures as they come to them. Typically, the comments relate to the wording of a particular part of the text; it is less usual to receive comments that relate to inconsistencies or interrelations between different parts of the standard.

While relatively little appears to have been published concerning review of standards, the author's experience does not seem uncommon, with other authors also bemoaning the quality of ad hoc standard review processes and seeking more rigorous methods (Graydon and Kelly 2013, Steele and Knight 2014).

Wong et al. (2014) did attempt to carry out a systematic review of five standards used in software safety. They scored them against twelve evaluation criteria that questioned how thoroughly each standard covers topics they deemed important like quality assurance and complexity management; if techniques like cost-benefit analysis or integrity levels were included; and other factors such as ease of use and active maintenance of the standard. They found some standards scored higher against some criteria and some against others, and suggested that projects should select their standards carefully to suit their needs. This analysis seems a little unsatisfactory: there was a justification for each criterion, but no explanation of how they chose the set as a whole. Their results seem less indicative of the quality of the standards, and more a consequence of the fact that the standards they evaluated had been written for different purposes, to fit into different regimes. Of the five standards evaluated, three were general system safety standards, rather than being software-specific (Def Stan 00-56[2], the Federal Aviation Administration System Safety Handbook and Mil-Std-882D[3]). One was not safety-specific (DO-178B, which addresses development of safety-critical software, but assumes the safety analysis will be performed and safety requirements set according to other standards). Of the five, only NASA-STD-8719.13B[4] was a dedicated software safety standard.

While software developers who have a free choice of safety standard may welcome some abstract criteria to aid their selection, standards developers need a different sort of criteria. They need to understand whether their particular standard is good for its intended purpose. Wong et al.'s work tells us that standards should be easy to use and have a good coverage of the topics deemed relevant to their scope. However, it does not give clear guidance on how to evaluate a standard on its own.

Graydon and Holloway (2015) also investigated the evaluation of software safety standards, motivated by the lack of evidence for their efficacy. They argued that there is little evidence to show that either the standards or the 'recipes' used to comply with them actually work. Without this, the apparent correlation reported between use of safety standards and lack of accidents could just be down to developers taking care when working with critical systems. Further, Graydon and Holloway claimed that there is rarely a testable hypothesis of what it means for a software safety standard to 'work'. In order to evaluate such a standard properly, one must first gain a clear understanding of what the standard is supposed to achieve and what the evaluation is expected to test, then plan accordingly.

The software engineering community often advocates various methods of Verification and Validation (V&V) to ensure the quality of software code (Ould 1999, Patton 2005). However, it can also be argued that the usefulness of V&V techniques extends beyond code to other artefacts used in the software development process. Taking this idea, we will explore how use of methods from software engineering, systems engineering and software safety can assist in evaluating and improving the quality of standards.

---

[2] The UK Ministry of Defence Safety management requirements for defence systems

[3] The US Department of Defense Standard practice for system safety.

[4] The NASA Software Safety Standard.

2.2 A review of reviews

2.2.1 Naïve Review

To provide a baseline for comparison, a 'naïve' review of Def Stan 00-055 Issue 4 was carried out by the author. This involved reading a hard copy version of the standard and marking up apparent issues in red pen (approx. 4.5 hours work), then re-reading and recording a description of each issue, with a proposed resolution, into a comments table (a further 6 hours). The impact of the identified issues were scored according to Table 1; issues were also classified into 17 types of problem.

Table 1. Issue impact descriptors.

Impact	Descriptor
High	Issues that appear to compromise the intent of the document.
Medium	Issues that affect the meaning of the document, but do not appear to compromise its intent.
Low	Incorrect, or makes the text harder to understand, but does not significantly affect the meaning of the document.
Readability	Issues of punctuation, grammar, style and similar that detract from the readability of the document, but are not otherwise incorrect.

170 issues were recorded, the majority being of Low impact or only affecting readability. Grammar, ambiguity and punctuation were the most common categories. An ironic example is the first paragraph of the Foreword. This contains a punctuation error, a spelling mistake and bad grammatical construction in the revision note. Instead of explaining that errors in the text have been fixed, it actually reads that the standard has been updated to "include ... minor grammatical and textural *[sic]* errors" which, while true, is presumably not what was intended! This example is interesting from two viewpoints: it illustrates that grammatical errors can be significant as they can change or invert the meaning of the text; and equally that they are not always worth fixing. The paragraph does not affect the requirements of the standard, and would inevitably have been replaced by a new revision note when the next issue is published, even had the mistakes not been spotted.

The issues identified with the greatest impact tended to be ambiguities (especially with regard to which requirements were mandatory) or omissions in the text.

2.2.2 ISO/IEC checklist review

Use of checklists is recommended in code inspections to prompt checks for omissions and avoid reviewers focusing just on what is there, rather than what is missing. Checklists also provide the ability to capture learning for experience (Fagan 1976, Ould 1999, Patton 2005). The International Organization for Standardization (ISO) and International Electrotechnical Commission (IEC) include a *Checklist for Writers and Editors* as an annex to their Directives (ISO/IEC 2021), which appears appropriate for review of an assurance standard.

Applying the ISO/IEC checklist to Def Stan 00-055 took less time than the naïve review (2.5 hours), but resulted in fewer issues being identified (59). However, over 90% of these were new issues, and the majority of the issues were of Medium impact. The most significant category of issues, both in terms of quantity and impact, related to confusion around which parts of the standard were normative and which informative.

2.2.3 Enhanced Checklist

Having used a checklist from the standards community, the author then looked to software engineering for inspiration to produce an improved checklist. The checklists used in a Fagan inspection are supposed to condition inspectors to seek high-occurrence, high-cost error types (Fagan 1976). The previously obtained results identify these types of error, theoretically allowing a more efficient checklist to be generated – we want to identify the common problems more easily and also reveal more significant problems that are harder to spot.

In practice, this is easier said than done. The most common problems (42% of the total) related to grammar, punctuation, omitted words and spelling; and it was not clear how a checklist would help. One might have hoped that these issues could be detected automatically, but the Microsoft Word spelling and grammar check did not reveal any of the issues found by the manual reviews (it had may well have already been used in the drafting process). Many of the other issues had already been found using the ISO/IEC checklist – how could this be improved further?

Taking inspiration from software requirements engineering, a list of positive quality attributes of assurance standards was compiled, drawing on suggestions from multiple sources (Ould 1999, Hull 2005, Patton 2005), as shown Table 2.

Table 2. Attributes of good requirements for standards

Abstraction	Consistency	Non-redundancy	Structure
Accuracy	Currency	Precision	Uniqueness
Atomicity	Feasibility	Relevance	Verifiability
Clarity	Legality	Satisfaction	
Completeness	Modularity		

These attributes were then applied as a checklist, both as general prompts for a read-through of the standard, and where possible, to inspire keyword searches for specific issues (e.g. finding 'and' to identify where requirements were non-atomic). Overall this process took approximately six hours and the issues it revealed were both more numerous and higher impact than using the ISO/IEC checklist. There was some overlap (10%) with previously identified issues, but the vast majority of the issues fell into the new categories listed in Table . For 7 of the 17 categories, no issues were found, hinting that there is perhaps room for refinement of the list.

2.2.4 Argument review

Ould maintains that the potential for V&V arises from formalism. He argues that the more structure and formality that is involved in creating a product, the more well-defined its meaning, and the easier it is to check (Ould 1999). A manual review of text appears to be an effective generic V&V technique that can be applied to standards, but these reviews are formal only in terms of their process, not the treatment of their underpinning semantics. Reviews of text-based documents are also not especially effective in uncovering high-level issues with strategy or design (Ould 1999). Some formality can be introduced through using inspection techniques with rules, criteria and checklists, but to gain greater V&V potential we somehow need to exploit the underlying structure of the standard. One approach is to use diagrams and abstraction to help identify high-level faults in strategy (Ould 1999). This points to reviewing some kind of abstract version of the standard, suggesting that a modelling approach might be useful.

Models of argument structures are often used to construct safety cases for software. Making the argument explicit is intended to help the author explain their safety case and let reviewers identify problems in its logic. Argument structures have also been used to model assurance standards (Ankrum and Kromholz 2005, Galloway et al. 2005), and as a basis for their review (Graydon and Kelly 2013). One can construct an argument to represent how the standard's aims are to be met, then use this structure as a drafting framework. The argument structure can be reviewed for completeness and consistency, then the standard can be verified against it. The author has used this method successfully to draft policy documents. Even when this method of drafting has not been used, a standard can be modelled retrospectively as an argument structure to facilitate review. This approach can help reviewers verify that the high-level goals of a software safety standard have been properly decomposed into requirements placed on the software. They can also check that meeting these low-level requirements will plausibly satisfy the overall goal. When using this approach, the need to set a top-level goal forces modellers to address the question of 'what it means for a standard to "work"' (Graydon and Holloway 2015).

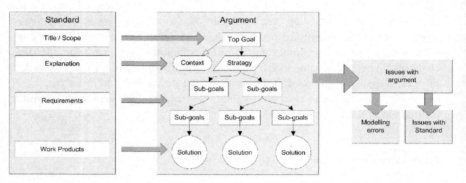

Fig. 3. Overview of argument modelling method.

To test out argument modelling as a review method, an argument was constructed using the Goal Structuring Notation (GSN) (ACWG 2018) in the ASCE tool (Adelard 2018), mapping parts of Def Stan 00-055 to GSN elements as shown in Fig. 3. This produced a complex network of 111 nodes, with a criss-cross of lines implying a tight coupling between different parts of the standard. Approximately half the issues found in the review were identified during the process of constructing the model, typically relating to requirements that were non-atomic requirements or had no obvious method of satisfaction. Further issues were found using the structure-checking tool built into ASCE. Manual review of the argument structure was carried out using guidance from (Hawkins and Kelly 2010), (Graydon and Kelly 2013), and the GSN Community Standard (ACWG 2018) – the latter document proving to provide the most practical advice. However, visual inspection of the structure did not reveal many further issues. Overall, the process took around ten hours, similar to the initial naïve review. A key high-impact findings was that various requirements that appear important to the standard are not actually well-defined.

Many of the results stem from the text of the standard being unclear or badly structured, in a way that makes it hard to model. The implication is that the text will also be hard to use in practice. The goal structure review focused on whether, if achieved, the requirements would satisfy the aim of the standard. However, it was less helpful in highlighting where those requirements might be practical to model but impractical to achieve.

2.2.5 Relationship modelling review

Goal structures appear to give useful insight into software safety assurance standards, but only present one view of the standard. Software design reviews are often supported by modelling that presents multiple coherent views of the software, using notations such as the Unified Modelling Language (UML). Different views can reveal potential problems with different types of relationship, such as temporal relationships between activities required by a standard. One framework developed specifically for use with assurance standards is the Reference Assurance Framework (RAF) metamodel (de la Vara et al. 2016).

Reviewing Def Stan 00-055 against the RAF metamodel identified 63 elements referred to in the standard that could be mapped to classes within the metamodel, including requirements, activities, roles and artefacts. The author attempted to construct a model using these classes in the Opencert toolset (Polarsys 2018), but found the functionality it provided impractical to use for a review. Attempts at constructing a process diagram to represent the activities required by the standard also failed, due to a lack of detail in Def Stan 00-055 about the sequencing of its activities. Instead, UML class diagrams were constructed, showing a static view of the types of relationship between entities described in the standard.

The results from this review were difficult to classify. Most of the issues identified by the other methods could be linked to specific portions of the text, with just a few general comments. They were also relatively easy to assign categories to. Modelling relationships identified fewer, but more far-reaching issues. They typically related to problems that affected several different parts of the text, often blending issues of consistency, completeness and clarity.

2.3 Filtering out problems

The baseline naïve review of Def Stan 00-055 Issue 4 revealed numerous issues. These were of a generally low significance, such as formatting, punctuation and style issues. Fixing these would have made the standard a little more readable but would not have made it much easier to understand or changed its meaning. However, these issues were not found by the other methods, and addressing them would have made the standard look more professional. This could be important from the point of view of acceptance of the document by its stakeholders. The naïve review also identified various more significant types of issue not found by the other methods, such as use of incorrect terms or outdated references, which depend on the expertise and knowledge of the reviewer. One might reasonably assume that other reviewers would find different issues, or might take a different view on their validity or relative importance. This implies that to increase the robustness of this review method, one should use multiple reviewers with different backgrounds and knowledge. One could also task them to approach the review from the viewpoint of different roles, as suggested by for code inspections (Fagan 1976, Patton 2005).

The two checklist-based reviews increased both efficiency and effectiveness – for those issues they addressed. They were quicker to conduct than thorough proofreading, but found more medium- and high-impact issues. While effective at finding issues in their scope, their focus on these issues meant other types of problem got overlooked. The ISO/IEC checklist mainly found issues with the presentation and structure of the standard, but the checklist based on software requirements engineering principles found more substantive issues with the standard's actual requirements.

The argument review produced interesting findings. Trying to identify the top-level goal revealed inconsistency in whether the standard was more about ensuring safety or 'design integrity' (freedom from flaws that might contribute to hazards). This recalls the discussion about needing a clear purpose for software safety standards (Graydon and Holloway 2013), and opens a debate about the difference between software correctness and safety.

Building models of the relationships between the concepts in Def Stan 00-055 was more difficult than anticipated. The problem was not that the method of modelling seemed unsuited. Indeed, the RAF metamodel provided a helpful way to think of the constituent elements of the standard's requirements, and the widespread use of UML in software engineering meant that easy-to-use modelling tools were widely available. Rather, the standard did not contain the anticipated information to support coherent models. While this review produced fewer specific issues than expected, the issues it did find were more fundamental, revealing several areas where what appeared to be important concepts in the standard were not addressed properly. It also highlighted that the standard ought to be written in a way that makes the relationships between its concepts more obvious. A possible way to do this would be to design models of the processes, obligations, interactions and other relationships described in the standard before starting to draft the next iteration.

Fig. 4 shows the relative quantity of issues revealed by the different review methods. It shows that the traditional method of naively reading a standard and commenting did indeed highlight more potential problems, but these were generally less important than those found by the more sophisticated methods. However, this hides the fact that the different methods tended to find different types of problem: each method was valuable in a different way.

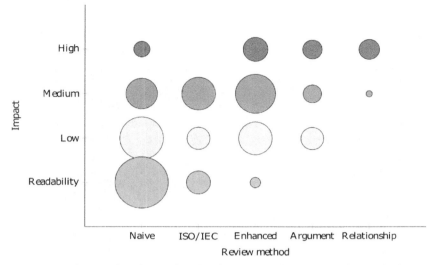

Fig. 4. Relative quantity of issues found broken down by impact and review method.

Graydon and Holloway (2015) suggested that one approach to understanding the intent of a software assurance standard is to consider it as part of a filter model, similar to that previously been proposed for regulation (Steele and Knight 2014). Graydon and Holloway saw standards as successful if they 'filter out' certain problems from software, either by making safety issues easier to spot, or by encouraging practices that reduce their likelihood or avoid them altogether.

We can adapt this filter model to the review of standards, by selecting review techniques to filter out different issues from the standard. As with a physical filter, it is more efficient to use coarse filters first: techniques that are likely to find major issues fast, rather than clog up the process with fine detail. Fig. 5 illustrates the principle, showing how different types of review checks can be used to filter out different types of issues, starting with those likely to require the most fundamental changes to put right if present in a standard.

In practice, it is likely to be impractical to apply all the different types of technique shown in Fig. 5 to a given review project. While the issues revealed through this work (Inge 2019) were included in the update of Def Stan 00-055 from Issue 4 to Issue 5, constraints on resource and timescales meant that it was not considered feasible to deploy the techniques described here more widely in the formal review.

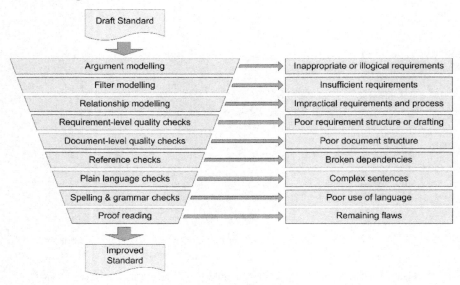

Fig. 5. A filter model for reviewing standards

Partly, this would have been a duplication of effort, and partly it was thought that construction of models and training reviewers in the use of the more sophisticated review methods would take disproportionate effort. However, use of these techniques might have more merit when designing a new assurance standard from scratch.

3 Conclusions

While it is a moot point whether software safety assurance standards are actually software, this research has shown that concepts from software engineering can be applied to make reviews of standards more effective. The requirements-setting parts of standards are similar enough to software requirements to make good practice from software requirements engineering applicable; and standards contain enough internal structure to make modelling methods useful. For both software and standards, there is a benefit to reducing complexity and making artefacts easier to understand, both to avoid errors and to aid maintenance.

However, there are some important differences. The use of natural language in standards makes it much harder to automate the review process, and the long iteration period of standards documents limits the return on investment provided by developing tools or learning involved techniques for their review. Typically assurance standards are only updated every few years, while agile software development may iterate versions every few weeks.

A potential area for both improving the drafting of standards and making them more amenable to review is the greater use of modelling in their construction (Model-Based Standards Engineering?) Representing standards using structured models could make automated checking more feasible, as well as potentially presenting different views of their requirements, that might reveal problems more easily to human reviewers.[5] However, to be cost effective, this kind of model would be likely to need to be built from the inception of the standard, rather than reverse engineered later.

A more practical way of improving reviews would be greater use of checklists as a prompt to reviewers to look for particular types of problem, and guide them as to how these might be found. Checklists based on desirable aspects of software or software requirements appear to have merit here, and could be applied more widely without undue cost.

3.1 Areas for further research

A key area for research would be automated reviews for standards. While some tools such as spelling and grammar checkers are available, standards writers have nothing to compare to the range of automated tools for testing and verification available to software developers. International standards organisations are researching machine-readable standards (Bielfeld and Rodier 2021); combining semantically marked-up standards with formal models using methods such as the Reference Assurance Framework (de la Vara et al. 2016) would help bring useful tools for standards review a step closer.

Due to its scope as an MSc project, the research discussed in this paper has been limited to review methods that can be accomplished by a solo reviewer, but there is scope to research the benefits of group methods. Fagan recommended four people as the optimum size for a software code review (Fagan 1976), and many of the more recent practices grouped under the "agile methods" banner are intended for use by small development teams. However, standards are intended for re-use on multiple projects and have a much broader range of stakeholders (and hence potential reviewers) than typical software code. The uplift of DO-178B to DO-178C involved 374 active participants, creating a tension between the need to build wide consensus and the need to draft a coherent document (Daniels 2011). The software industry has developed distributed collaboration tools such as Bugzilla and Jira for issue management, and GitHub and SourceForge for source control; there is scope to investigate whether similar tools and methodology could help coordinate input from participants in large standards reviews.

---

[5] A model-based approach to standards might also prompt more re-use of common patterns or templates. This could perhaps help spread good practice and increase consistency between standards, aiding both reviewers and users of standards .

Finally, this work has focused mainly on the 'non-functional' aspects of standards: qualities such as self-consistency or readability. More research is needed to verify that safety assurance standards actually have the desired impact in terms of improving safety.

Acknowledgments The author is grateful to his supervisor Peter Bloodsworth and the Software Engineering Programme at the University of Oxford for their support in the MSc research on which this paper is based, and to Defence Equipment & Support for funding the MSc through their upskilling fund.

The Chinook image at Fig. 1 is provided by Defence Imagery / SAC Mark Parkinson and used under the Open Government Licence: https://www.nationalarchives.gov.uk/doc/open-government-licence/version/3/

Disclaimers Views expressed in this paper are those of the author, and not necessarily those of his employer.

References

ACWG (2018) GSN Community Standard version 2, Assurance Case Working Group. https://scsc.uk/scsc-141B. Accessed 17 September 2022.
Adelard (2018), ASCE software, https://www.adelard.com/asce/choosing-asce/index/. Accessed 18 June 2018.
Ankrum TS and Kromholz AH (2005) Structured assurance cases: Three common standards. In Ninth IEEE International Symposium on High-Assurance Systems Engineering (HASE'05). Institute of Electrical and Electronics Engineers (IEEE). DOI: 10.1109/hase.2005.20.
Bieldfeld A and Rodier K (2021) What's next in Standards and Standards Publishing at ISO and IEC. In Typefi Standards Symposium 2021. https://www.typefi.com/standards-symposium-2021/whats-next-in-standards-publishing-iso-iec/. Accessed 15 November 2022
Bourn SJ (2004) Battlefield helicopters. House of Commons report HC 486 2003–2004. National Audit Office. https://www.nao.org.uk/wp-content/uploads/2004/04/0304486.pdf. Accessed 29 May 18
Burr T (2008) Chinook Mk3 helicopters. House of Commons report HC 512 2007–2008. National Audit Office. https://www.nao.org.uk/wp-content/uploads/2008/06/0708512.pdf. Accessed 29 May 18
Daniels D (2011) Thoughts from the DO-178C committee. In Proceedings of the 6th IET International Conference on System Safety 2011. Institution of Engineering and Technology. DOI: 10.1049/cp.2011.0266
De la Vara JL, Ruiz A, Attwood K, Espinoza H, Panesar-Walawege RK, López Á, Del Río I and Kelly T (2016) Model-based specification of safety compliance needs for critical systems: A holistic generic metamodel. Information and Software Technology, vol. 72, pp. 16–30. DOI: 10.1016/j.infsof.2015.11.008.
Fagan ME (1976) Design and code inspections to reduce errors in program development. IBM Systems Journal, vol. 15, no. 3, pp. 182–211. DOI: 10.1147/sj.153.0182
Galloway A, Paige R, Tudor N, Weaver R, Toyn I and McDermid J (2005) Proof vs testing in the context of safety standards. In 24th Digital Avionics Systems Conference, IEEE. DOI: 10.1109/dasc.2005.1563405

Graydon PJ and Holloway CM (2015) Planning the unplanned experiment: Assessing the efficacy of standards for safety-critical software. Technical Memorandum NASA/TM-2015-218804. NASA Langley Research Center.

Graydon PJ and Kelly TP (2013) Using argumentation to evaluate software assurance standards. Information and Software Technology, vol. 55 no 9 pp. 1551–1562. DOI: 10.1016/j.infsof.2013.02.008.

Hawkins RD and Kelly TP (2010) A systematic approach for developing software safety arguments. Journal of System Safety, vol. 46, no. 4, pp. 25–33. ISSN: 0743-8826.

Hull E, Jackson L and Dick J (2005) Requirements Engineering. Springer-Verlag. ISBN: 978-1-85233-879-4. DOI: 10.1007/b138335

IEC (2010) IEC 61508 series – Functional safety of electrical/electronic/programmable electronic safety-related systems. International Electrotechnical Commission

Inge JR (2019) Improved Methods for Review of Software Assurance Standards using Def Stan 00-055 as a Case Study. University of Oxford

ISO/IEC (2021) ISO/IEC Directives, Part 2:2021 – Principles and rules for the structure and drafting of ISO and IEC documents. International Organization for Standardization, International Electrotechnical Commission. https://www.iec.ch/news-resources/reference-material. Accessed 17 September 2022

MOD (2016) Def Stan 00-055 Issue 4 – Requirements for safety of Programmable Elements (PE) in defence systems. Ministry of Defence

MOD (2021) Def Stan 00-055 Issue 5 – Requirements for Safety of Programmable Elements (PE) in Defence Systems. Ministry of Defence

Morse A (2013) Major projects report 2012. House of Commons report HC 684-I 2012–13. National Audit Office. https://www.nao.org.uk/wp-content/uploads/2013/03/Major-Projectsfull-report-Vol-1.pdf. Accessed 29 May 18

Ould MA (1999) Managing software quality and business risk. John Wiley & Sons Ltd. ISBN 047199782X

Patton R (2005) Software testing. Sams Publishing. ISBN 067232798-8

PolarSys (2018) Opencert website. https://www.polarsys.org/opencert/. Accessed 10 January 2019

Steele P and Knight K (2014) Analysis of critical systems certification. In 2014 IEEE 15th International Symposium on High-Assurance Systems Engineering, pp. 129–136. DOI: 10.1109/HASE.2014.26

Wong WE, Gidvani T, Lopez A, Gao R and Horn M (2014) Evaluating software safety standards: A systematic review and comparison. In 2014 IEEE Eighth International Conference on Software Security and Reliability-Companion. pp. 78–87. DOI: 10.1109/SERE-C.2014.25

IEC 63187 – Tackling complexity in defence systems to ensure safety

James Inge

Defence Equipment & Support

Bristol, UK

Phil Williams

Engineer for Safety Limited

Hastings, UK

Abstract *IEC 63187 is the new functional safety framework being developed by the International Electrotechnical Commission for the defence sector. In this sector, applications are typically complex systems, elements of which may themselves be both technically complex and managerially complex systems in their own right: developed by different suppliers, to different standards, and at different stages in their product lifecycles. Defence systems are also subject to dynamic changes of risk, depending on the context of their deployment. Existing safety standards are not well adapted to this level of complexity. They tend to be aimed at single organisations rather than complex hierarchies, and to focus on the failures of system elements, rather than important emergent properties of the overall system. The new international standard in development, IEC 63187, tackles these problems using modern systems engineering principles. It applies the ISO/IEC 15288 life cycle processes to supplement IEC 61508 and other safety standards, proposing an approach that allows requirements to be tailored to the risk and managed across multiple system layers. This framework is designed to be open, for compatibility with different national approaches to assurance and risk acceptance, and with different traditional standards for realisation of individual system elements. This paper discusses the motivation, principles and approach of IEC 63187 and gives an update of the progress of the drafting of the document through the standardization process.*

© James Inge, Phil Williams 2023.
Published by the Safety-Critical Systems Club. All Rights Reserved.

1 Introduction

The International Electrotechnical Commission (IEC) is currently drafting a new international standard titled 'Functional Safety – Framework for safety critical E/E/PE systems for defence industry applications'. 'E/E/PE' relates to Electrical, Electronic and Programmable Electronic systems, including software and complex electronic hardware. Such systems are increasingly prevalent in defence applications, even in roles where mechanical systems have traditionally been used. IEC 63187 aims to help suppliers demonstrate that complex defence products, systems and services incorporating E/E/PE are acceptably safe for their customers to operate. This paper explains why a new international standard is necessary in this area, and introduces some of the key innovations in its approach.

1.1 The challenge of defence systems

Systems in the defence sector often have characteristics that are not well catered for by existing functional safety standards:

Managerial complexity: Defence applications are often 'systems of systems' in several of the senses used in the Systems Engineering Body of Knowledge, in that the system elements that make them up are separately defined, acquired and integrated (SEBoK 2021). These elements may be a combination of bespoke new developments, off-the-shelf components, customisations of existing designs, and 'legacy' equipment that is already in service. The different elements are often specified and procured separately from different suppliers at different times, and increasingly may be supplied as services rather than traditionally acquired hardware. Hence they may be at different stages in their product life cycles when brought together to deliver an overarching capability. Existing functional safety standards tend to be limited in scope, and are often intended to be applied within a single organisation, rather than across a complex supply chain.

Technical complexity: Major defence capabilities are often made up of a number of system elements that are complex systems in their own right. Since the 1960s, systems engineering techniques have been developed to manage this complexity, in defence and other industries. However, current functional safety standards do not necessarily apply systems engineering principles and anticipate recursive application through a hierarchy of systems. Safety Integrity Levels (SILs) and similar concepts become difficult to apply in complex systems hierarchies: it becomes hard to decompose SILs and assign them over multiple layers of the hierarchy, especially when the different system elements may be managed separately.

There is also a tendency for standards to focus mainly on guaranteeing safety by controlling the impact of failures of individual system elements. However, in complex systems, emergent properties are a concern, and it is possible for systems to behave unsafely without failures of their individual elements.

Dynamic risk: The hazards and potential losses involved in military systems are dependent on the context of operation, and there is a balance to be made between the safety and the capability of the system. While this is true of most systems, the operating context for military systems can change frequently and rapidly during their operation, resulting in changes to safety objectives and trade-offs. For example, changes to the threat posed by hostile actors may mean that it is necessary to compromise some safety objectives in order to complete the mission. There is often an assumption in functional safety standards that the level of risk will remain largely constant.

Customer determination of risk acceptability: in many other industries, the acceptable level of risk is determined to a certain extent by civil regulation; or the organisation supplying the product is able to set their own risk appetite. In defence, often the arbiter of risk acceptability is the organisation acquiring the system, normally a national defence ministry or an agency working on their behalf. Civil safety legislation often explicitly excludes defence systems from its scope, or gives powers to the government to exempt particular applications in the interests of national security. Functions performed by defence systems are sometimes also uniquely military in nature, and not well covered by civil product safety standards. Defence procurement organisations have a dilemma: they are generally held accountable by their government, so need assurance from suppliers that the equipment they procure will be safe to operate. However, they do not wish to overly constrain implementation options, limit operational capability or impose unnecessary costs on their projects.

While all of these characteristics are common in systems found in the defence sector, in practice they could be found in other sectors using complex technology or where the interactions of system elements is complex. IEC 63187 is initially directed at the defence sector; however, there is nothing inherently defence specific in its normative requirements.

1.2 The IEC 63187 approach

When developing a complex system, safety is neither something that can be managed independently, nor the end goal of the system. It is an emergent property of the system as a whole, rather than a separate feature that can be designed in. Similarly, safety is not the outcome of a single technical or managerial process, but the result of the multi-disciplinary combination of activities that go into designing, manufacturing, deploying and operating the system.

IEC 63187 recognises this, and also recognises that a standardized body of good practice already exists in disciplines like systems engineering, risk and quality management, which may not be specifically aimed at managing safety, but nonetheless supports delivery of safe socio-technical systems. Rather than attempting to duplicate these standards, IEC 63187 builds on them to explain how they can be extended using systems thinking and systems engineering to produce an effective framework for managing the functional safety aspects of a complex system.

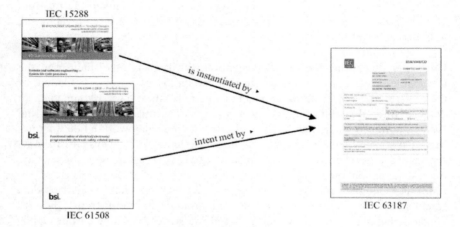

Fig. 1. IEC 63187 pedigree

In particular, IEC 63187 builds on the systems engineering framework of IEC 15288 (IEC 2015). Although IEC 61508 is the 'horizontal standard' or 'Basic Safety Publication' for functional safety of E/E/PE systems[1], the detailed approach described in IEC 61508 is only appropriate for those defence systems that fit the functional concept described in the standard. Instead of building directly on IEC 61508, IEC 63187 aligns more directly to IEC 15288. It takes the concept of systems engineering processes managed within a life cycle framework, and specifies additional requirements on those processes to achieve the intent of IEC 61508 for defence systems. These additional requirements are targeted at ensuring both that the safety objectives for the system will be achieved, and that adequate assurance information will be produced to give the acquiring organisation

---

[1] Meaning that it gives "fundamental principles, concepts, terminology or technical characteristics, relevant to a number of technical committees and of crucial importance to ensure the coherence of the corpus of standardization documents" (IEC 2022a)

confidence that this is so. Beyond this, IEC 63187 also provides a framework for understanding the interaction between hazards at different layers of the systems hierarchy, and specifying safety requirements on the lower layers.

IEC 63187 does not specify functional safety requirements for the development or realisation of particular system elements. However, it puts in place a framework by which their requirements and safety objectives can be derived. IEC 61508 can still be used under IEC 63187 to realise those system elements for which it is suited. Similarly other standards such ISO 26262[2] or DO-178C[3] could be used, as appropriate to the application domain.

---

[2] ISO 26262: Road Vehicles – Functional Safety.

[3] DO-178C: Software Considerations in Airborne Systems and Equipment Certification.

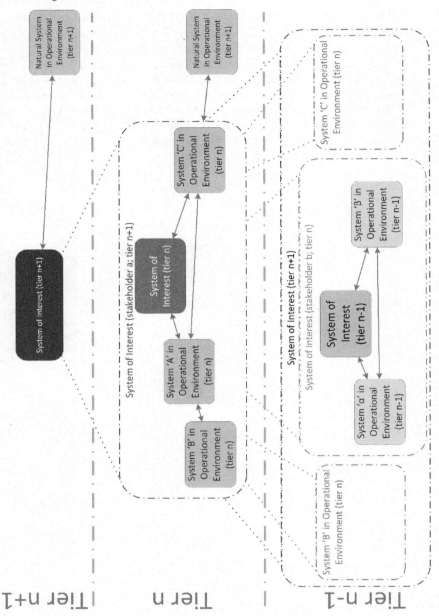

Fig. 2. Hierarchy of System Tiers (IEC 2022)

2 How does IEC 63187 tackle complex systems?

2.1 Recursion and iteration through the systems hierarchy

While traditional standards assume a fixed hierarchy, are intended to be applied to a complete system and have different requirements for different system elements such as hardware and software, IEC 63187 explicitly recognises an abstract, flexible, systems hierarchy. As shown in Fig. 2, at each tier of the hierarchy, there is a bounded 'system of interest' operating within a certain environment[4]. The environment is outside the scope of the engineering control for the system of interest. If aspects of the environment do need to be engineered, then a higher tier can be added to the hierarchy with those aspects included in the scope of the higher-tier system of interest. The system elements composing a system of interest can either be considered as atomic units that can be realised directly and do not need further analysis, or they could be considered as systems in their own right, and analysed in a lower tier in the hierarchy. The system of interest forms part of the operating environment for systems in the tier below. This hierarchic approach allows the management of complexity by allowing the detailed design of individual system elements to be abstracted, allowing analysis at a higher level. This approach allows systems to be considered at a high-level tier that are in the operational domain and inclusive of people, aspects of the natural environment and technological systems.

IEC 63187 is intended to be applied recursively throughout the hierarchy until the bottom tier, where more specific requirements can be set for realisation of particular system elements. Depending on the systems breakdown and supply chain involved, individual participants may apply the standard at multiple tiers, or just one. This approach allows systems to be considered at differing levels of abstraction, and of aggregation of disparate physical elements. These facilitate the use of the standard from early concept stage through to in-service operation, and beyond. They also allow for the standard to be applied throughout the supply chain from a user with the need for a capability, through its acquisition agency right through to suppliers of system elements.

---

[4] In practice there can be many systems of interest at each tier, each of which may have a different 'owner' of that interest.

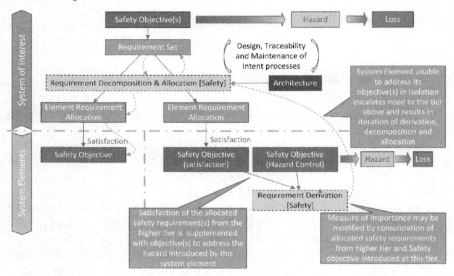

Fig. 3. Derivation of safety objectives and requirements (IEC 2022b).

At each tier of the system, safety objectives for the system element of interest are expected to be set to allow it to satisfy safety requirements set by the tier above, as shown in Fig. 3. In turn, that tier will set safety requirements to be met by the objectives of lower-tier system elements. In this way, requirements are derived for the bottom-tier system elements that can be traced back to achievement of the top-level functional safety objectives for the overall system.

New hazards can also be introduced at any system layer. They could result from failure modes of systems elements, deliberate implementation choices, or unintentional interactions between system elements. Such hazards may well not be present in lower-tier system elements, but only emerge through integration. IEC 63187 requires analysis to take place to reveal whether such hazards are present and further safety objectives to be set to control them. In some cases, these hazard control objectives may be discharged by setting safety requirements on lower-tier system elements. In others, this will not be feasible, and it will be necessary to iterate the requirement setting activity for the tier above. This may result in additional safety requirements being placed on other system elements, or even generate a need for a new system element to control the hazard. In this way, IEC 63187 seeks to address emergent hazards.

2.2 Risk Model

IEC 63187 adopts the ISO/IEC/IEEE 15288 view of risk as 'the effect of uncertainty on stakeholder objectives'. It does not use the traditional measure of risk as a function of the likelihood and severity of an outcome as this is not necessarily helpful in the context of safety analysis in an abstract systems hierarchy.[5] The likelihood/severity approach also does not lend itself to dynamic risk scenarios, where the probability and severity can be expected to change more frequently than the analysis can be carried out. Instead, IEC 63187 focuses on uncertainty in the control of hazards, which are defined as system states or sets of conditions that, together with a particular set of environmental conditions, will lead to harm. Hazards are 'owned' and managed at the tier of the systems hierarchy in which they are necessarily introduced, for example by the choice of a particular technology to implement a system element. Where a hazard is identified, safety objectives are set to control the impact of this hazard and prevent it resulting in harm or loss. Requirements are then set and allocated to system elements to ensure that the safety objective is met. This approach lends itself to application of control theory and systems engineering-based techniques such as System-Theoretic Process Analysis (STPA) (Leveson and Thomas 2018).

Aside from being used to judge the tolerability of potential accidents, traditional standards use the likelihood/severity risk metric to define the level of rigour required in designing particular parts of a system, or the level of confidence required that particular requirements have been achieved. As IEC 63187 does not use this risk metric, it has to propose an alternative method to determine where effort should be prioritised to control hazards and provide assurance. To do this, it introduces the concept of a 'measure of importance'.

2.3 Measures of Importance

The IEC 63187 concept of a 'Measure of Importance' (MOI) describes the degree of confidence required when ensuring or assuring safety. It plays a similar role to concepts like Safety Integrity Levels (SILs) or Design Assurance Levels

---

[5] For example, it is not possible to assess the risk due to failure of a subsystem such as an electronic control unit (ECU) as an isolated system: it is necessary to understand the rest of the system in which the ECU operates, like a vehicle or aircraft, to understand the likelihood that failure of the ECU could propagate to a hazardous state in the top-level system. Further information is needed about the operating environment to understand the likelihood that an accident might result, and the severity of the harm caused. Such an analysis may be feasible in relatively simple systems using techniques such as Failure Modes and Effects Criticality Analysis (FMECA), but it is not feasible in more complex systems, where system elements are being independently developed and information about the higher system tiers is not available.

(DALs) in other standards, in defining the level of rigour to be applied in different systems engineering processes. The MOI concept is however more flexibly defined, to enable it to be recursively applied at different system layers. In fact, although it provides an example in an informative annex, IEC 63187 does not define a specific MOI schema, but requires one to be drawn up as part of the safety acceptance strategy and agreed between the acquirer and supplier. This allows the concept to be tailored to align to national legislation or regulatory requirements, and to reflect particular concerns of the acquirer. For instance, the schema can prioritise harm to humans as more important than financial loss, seek extra rigour for particular types of hazard that cause societal concern (e.g. radiological hazards), or require extra scrutiny for particular technologies.

Measures of importance can be applied to hazards, safety objectives and requirements, potentially with different scales for each. The MOI for a hazard will be based on the severity of the associated loss, conditioned by factors such as the organisation's risk appetite in different operational contexts. While likelihood of the loss would but not be taken into account directly, the degree of contribution of the hazard to the loss could also be a conditioning factor[6]. Hazard MOIs are used to set MOIs for associated safety objectives, which in turn are used to set MOIs for their supporting safety requirements, again with conditioning factors taken into account. These conditioning factors allow the MOI schema to reflect the overall safety strategy for the system, trade-offs between safety, capability and other concerns, and the importance of different system elements to the overall architecture. The allocation of MOIs to safety requirements means that there will be a flow down to lower-level system tiers. However, a translation may be necessary, as these tiers may use different MOI schemas. At the bottom system layer, there will also need to be a translation from the MOI schema to measures specific to the chosen implementation standards, such as SILs or DALs.

MOIs are a powerful and flexible concept, but have the potential to be confusing to use in practice. If MOI schemas are not set up appropriately, then application of IEC 63187 may not result in the acquirer gaining the assurance of safety that they desire. This should not be an insurmountable challenge. Acquirers already have to set their expectations for the level of assurance provided by their supply chain, but IEC 63187 makes the requirement more explicit. However, the success of the standard in this respect may well rest on the strength of the guidance available to help implementers to define practical MOI schemas.

---

[6] 'Conditioning factors' are factors that may influence the allocation of a measure of importance, to allow a higher or lower MOI to be allocated in particular cases. IEC 63187 does not define a particular set of conditioning factors, but allows the organisation using the standard to define them as part of their MOI schema. They might include factors such as the type of people at risk (civilian / military / enemy), the type of operational scenario involved (training vs military operations), or the degree of contribution of a safety objective to the overall safety architecture.

3 Relationship to UK Defence Standards

As an international standard, IEC 63187 needs to be capable of application in any country. This means that it has to remain independent of the requirements of particular legislative or regulatory jurisdictions or acquisition regimes, so will not reference particular national defence standards. It will also not necessarily align directly with the vocabulary in use in different countries, since this varies and common terms like 'hazard' can be interpreted differently, even in countries that share the English language (McDermid 2007). Instead, it will build on the common vocabulary used in other IEC and International Organization for Standardization (ISO) standards. However, development of IEC 63187 has been informed by knowledge of various national defence standards and the thinking behind them.

Notably, the conformance requirements of IEC 63187 have been derived from the same software safety assurance principles originally developed by (Hawkins et al. 2013), which feature as programmable element safety requirement principles in Def Stan 00-055 and Def Stan 00-056. This means that IEC 63187 takes a similar approach to assurance to the aforementioned Def Stans, and to other material based on similar principles, such as the Service Assurance Guidance (SAWG 2022). IEC 63187 also includes the concept of a 'safety case', albeit as a placeholder for all the information generated over the system lifecycle to show satisfaction of the standard. While not calling explicitly for a safety argument, the standard requires various claims to be documented in the safety case, and requires the acquirer and supplier to agree a safety acceptance strategy. This strategy allows the flexibility to specify the need for an explicit safety argument, or other nation-specific assurance requirements. This is intended to allow IEC 63187 to remain compatible with the UK's Def Stan 00-055, the US Mil-Std-882E, and other nation's safety management standards. It can also support the philosophy that only the Duty Holder responsible for operating a system safely is positioned to make claims about the overall safety of the deployed system, taking into account the operational environment and other lines of development, such as training or doctrine. In this context, the information provided through application of IEC 63187 does not provide the overall safety case itself, but supports the overall safety case made by the Duty Holder, when combined with arguments from other areas of their safety management system.

As the UK Ministry of Defence (MOD) has a policy of selecting civilian standards wherever practicable and military standards only where necessary, and prefers international standards to national or military ones (MOD 2022), there is likely to be interest in assessing whether IEC 63187 could replace Def Stan 00-055 and 00-056. While it is hoped that IEC 63187 will provide a convenient means to demonstrate compliance with those standards for complex systems, it is unlikely to replace them. As it only covers functional safety, IEC 63187 does not cover the complete scope of Def Stan 00-056. And for some less complex systems, it may be more appropriate to continue using implementation standards such as IEC 61508 directly.

For military systems within the scope of IEC 63187, there is still a compelling reason to retain the use of Def Stans: IEC 63187 has various requirements for the acquirer to define the interface between it and the supplier. This includes specifying the acquirer's requirements for a safety strategy and safety acceptance strategy (including the MOI schema), any particular methods or techniques the supplier is required to apply, and the safety artefacts they are to deliver. It also includes reaching agreement with the supplier on issues such as the MOI schema to be applied, or the compliance routes for already-realised system elements. Some of these points are project-specific, but others can be generic to a particular acquirer. For instance, internal regulations in a defence ministry may require particular safety artefacts to be generated. For this reason, the MOD is likely to wish to retain the use of Def Stans in some form, to standardize its approach to meeting these IEC 63187 requirements.

IEC 63187 could also be applied at tiers higher than that at which MOD procures systems, as part of understanding and managing the emergent interactions between systems it procures, or as part of studying the end user's needs and selecting suitable new procurement items to fit alongside existing systems to deliver the required capability.

4 Development progress

IEC 63187 is being developed under IEC Technical Committee TC65 (Industrial-process measurement, control and automation), by Subcommittee SC65A – Systems Aspects, the same part of the IEC that maintains IEC 61508. The Working Group drafting IEC 63187 (IEC SC65A WG18) has been meeting since 2018 and currently includes representatives from ten nations. It is drafting the standard in two parts. IEC 63187-1 will contain the normative parts of the international standard, along with informative material including an annex on the concepts and rationale of the standard. Further guidance will be provided in IEC TR 63187-2.

At the time of writing, IEC 63187-1 has been circulated as a Committee Draft (CD) for comments by IEC Members, i.e. national standards committees. The draft will be updated during 2023 based on the comments received, and is planned to be circulated as a CD for an approval vote (CDV) in early 2024. Assuming that the CDV is approved but further technical comments are made, it is likely to be issued as a Final Draft International Standard (FDIS) in late 2024 and eventually published in 2025.

Drafting has started on the supporting guidance in IEC TR 63187-2. This part of the standard will have the status of a Technical Report rather than a full International Standard, meaning that it is entirely informative, rather than setting any normative requirements. Technical Reports have a more flexible approval route, meaning that there is scope to shape the Part 2 guidance to address comments raised against Part 1 of the standard, and still publish both parts at the same time.

Disclaimer Views expressed in this paper are those of the authors and not necessarily those of the Ministry of Defence or the International Electrotechnical Commission.

References

Hawkins RD, Habli I, Kelly TP (2013) The principles of software safety assurance. In Proceedings of the 31st International System Safety Conference. International System Safety Society. Available: https://www-users.cs.york.ac.uk/rhawkins/papers/HawkinsISSC13.pdf. Accessed 3 September 2022

IEC (2010) IEC 61508 series – Functional safety of electrical/electronic/programmable electronic safety-related systems. International Electrotechnical Commission

IEC (2015) ISO/IEC/IEEE 15288:2015 – System and Software Engineering – System life cycle processes. International Electrotechnical Commission

IEC (2022) Committee Draft 65A/1048/CD. Functional safety – Part 1: Framework for safety critical E/E/PE systems for defence industry applications. International Electrotechnical Commission

IEC (2022a) Horizontal Standards. International Electrotechnical Commission. https://www.iec.ch/news-resources/horizontal-standards. Accessed 2 October 2022

IEC (2022b) IEC 63187-1 Functional safety – Framework for safety critical E/E/PE systems for defence industry applications – General presentation for CD circulation. TC65/SC65A/WG18. International Electrotechnical Commission

Leveson NG, Thomas JP (2018) STPA Handbook. https://psas.scripts.mit.edu/home/get_file.php?name=STPA_handbook.pdf. Accessed 17 September 2022

McDermid JA (2007) Comparison of MilStd 882E and Interim Defence Standard 00-56 Issue 3. In Gonzalez AM (Ed.), Proceedings of the 25th International System Safety Conference

MOD (2022) JSP 920 – MOD Standardization Management Policy – Part 1: Directive, Ministry of Defence. https://www.dstan.mod.uk/policy/JSP920_Part1.pdf. Accessed 3 September 2022

SAWG (2022) SCSC-156B – Service Assurance Guidance version 3.0. Safety Critical Systems Club Service Assurance Working Group. https://scsc.uk/r156B:1

SEBoK contributors (2021), System of Systems (SoS) (glossary), SEBoK. https://www.sebokwiki.org/w/index.php?title=System_of_Systems_(SoS)_(glossary)&oldid=65565. Accessed 2 October 2022.

Anticipating Accidents through Reasoned Simulation

Craig Innes[1], Andrew Ireland[2], Yuhui Lin[2] and Subramanian Ramamoorthy[1]

[1]School of Informatics, University of Edinburgh

[2]School of Mathematical and Computer Sciences, Heriot-Watt University

Abstract *A key goal of the System-Theoretic Process Analysis (STPA) hazard analysis technique is the identification of loss scenarios – causal factors that could potentially lead to an accident. We propose an approach that aims to assist engineers in identifying potential loss scenarios that are associated with flawed assumptions about a system's intended operational environment. Our approach combines aspects of STPA with formal modelling and simulation. Currently we are at a proof-of-concept stage and illustrate the approach using a case study based upon a simple car door locking system. In terms of the formal modelling, we use Extended Logic Programming (ELP) and on the simulation side, we use the CARLA simulator for autonomous driving. We make use of the problem frames approach to requirements engineering to bridge between the informal aspects of STPA and our formal modelling.*

1 Introduction

System-Theoretic Process Analysis (STPA) (Leveson, 2011) extends traditional hazard analysis techniques to include a range of causal factors that can affect system safety. A key goal of STPA is the identification of causal factors that give rise to unsafe control actions, which in turn could lead to hazards and ultimately an accident – what Nancy Leveson called a *loss scenario*. We propose an approach that aims to assist engineers in identifying potential loss scenarios. Our approach combines aspects of STPA with formal modelling and simulation. STPA is relatively informal in comparison to formal modelling. To bridge between the informal and the formal we use the notion of a problem frame (Jackson, 2001). The problem frame representation allows us to verify the consistency between *system-level constraints* and *control constraints* – which are both by-products of the STPA technique.

© University of Edinburgh and Heriot-Watt University 2023.
Published by the Safety-Critical Systems Club. All Rights Reserved.

The use of formal modelling also enables us to explore the assumptions an engineer has made about a system's intended operational environment. Specifically, it allows us to explore what loss scenarios may occur if such assumptions are flawed. The resulting loss scenarios, while abstract, can be used to tailor simulations. Through simulation we are able to identify more concrete and accessible loss scenarios. Our proposed approach is at the proof-of-concept stage. We illustrate it using a case study based upon a simple car door locking system. In terms of the formal modelling, we use Extended Logic Programming (ELP) (Gelfond & Lifschitz, 1991) and on the simulation side, we use the CARLA (Car Learning to Act) simulator for autonomous driving (Dosovitskiy, et al., 2017). In section 2, we provide background on STPA and the aspects of ELP and problem frames that we make use of in this paper. We also provide an overview of the CARLA simulator and the features that we rely upon in this paper.

2 Background

We follow the 4-step description of STPA given in (Leveson & Thomas, 2018). The first step involves identifying the potential *losses* (L), system-level *hazards* (H) and *system-level constraints* (SC). A hazard denotes a system state coupled with environmental conditions that lead to a potential loss. The role of the system-level constraints is to guard against the identified hazards from occurring. Step 2 focuses on the achievement of the system-level constraints. First, it involves defining a *control structure* – a diagrammatic description of a system's controllers and controlled processes, together with how they interact, i.e., *control actions* and *feedback*. Second, the responsibilities for each controller are defined, where a *responsibility* (R) represents a refinement of a system-level constraint. Each responsibility is identified with part of a controller's *process model* and *control logic*, which are informed by feedback from the processes that it controls. Finally, a control action is associated with each responsibility. Step 3 identifies how inadequate control could lead to a hazard by considering the following cases:

1. A control action that when not applied causes a hazard.
2. A control action that when applied causes a hazard.
3. A control action that when applied too early, too late or out of sequence causes a hazard.
4. A control action that is stopped too soon or applied too long causes a hazard.

Note that the final case is applicable to non-discrete control actions. For each unsafe control action, a corresponding *control constraint* is defined, which specifies a property that must be satisfied in order to prevent the controller exhibiting the unsafe control actions. Figure 1 summarises the first 3-steps of STPA. Despite design-level verification, the fourth step focuses on identifying how unsafe control actions could occur and lead to the violation of system-level constraints. As noted above, these are referred to as loss scenarios. In (Leveson, 2011), the process of identifying a loss scenario is described as '*the usual "magic" one that creates the contents of a fault tree.*' There are many ways in which unsafe control actions could occur. In section 4.4, we focus on *flawed assumptions* as the source of loss scenarios.

As mentioned above, we use the notion of a *problem frame* (Jackson, 2001) to bridge between the informal and the formal. A problem frame provides a way of verifying the consistency between a system-wide requirement (i.e., system-level constraint) and the specification of a software component (i.e., control constraint). The problem frame representation has both graphical and logical elements. Substituting the STPA control structure with a problem frame involves formalizing the domain assumptions associated with the system and its operational environment. An informal verification is represented by means of an *informal argument diagram* (Seater, et al., 2007) which will be illustrated in sections 4.3 and 4.4. In terms of formal logic, the problem frames approach is relatively agnostic. Our case study is quite simple, so we opt for classical propositional logic - specifically we use *Extended Logic Programs* (ELP) (Gelfond & Lifschitz, 1991). As a consequence, we ignore delays between observations and actions. More expressive formalisms could be used. ELP extends the normal notion of a Logic Program (LP), such as Prolog, with explicit negation (-). Similar to LP, a program in ELP is a collection of clauses in the following format:

$$L :\text{-} A_1, ... A_m, \text{not } B_1, ... \text{not } B_n.$$

while it allows literals in the body, e.g. A_i and B_j to be negated, e.g. $-A_i$ and $-B_j$.

Formal modelling is suited to relatively abstract models where the state space is small and can be exhaustively explored. In contrast, simulation allows one to selectively explore large state spaces. In terms of simulations, we have used *CARLA* (Dosovitskiy, et al., 2017), which is a simulation platform for testing the performance of automated vehicle systems. It provides a rich set of features for capturing the many complex interactions between the many integrated modules within a modern Autonomous Vehicle (AV) pipeline. In terms of physical dynamics and control, CARLA can simulate vehicles that are affected by a highly granular set of parameters. This can be as broad as overall mass, all the way down to gear ratio, drag coefficient of the chassis, tyre friction, and wheel stiffness. Also, CARLA provides libraries, e.g., Python API, to specify/program scenarios for both local control and global route planning of large numbers of vehicles on the road simultaneously.

A simulator provides a way to efficiently generate many thousands of example rollouts/scenarios of the AV system given a user-specified starting specification.

So, while symbolic domain knowledge may provide heuristics about which configuration parameters are likely to be relevant to safety, testing variants of those parameters out in simulation allows us to see how they affect behaviour in practice, and thus search for falsifying examples.

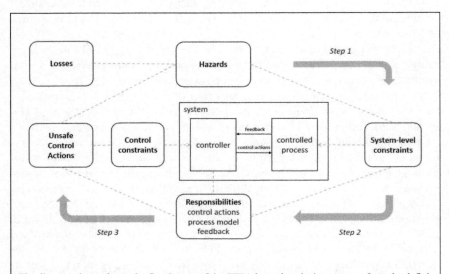

The diagram above shows the first 3 steps of the STPA hazard analysis process – from the definition of losses and hazards through to defining control actions and identifying how unsafe control actions can lead to hazards.

Fig. 1. Summary of steps 1-3 of STPA.

3 Related work

As illustrated above, the problem frame approach provides a technique for verifying the correctness of a system-wide requirement with respect to a specification and domain assumptions. In (Seater, et al., 2007) Alloy (Jackson, 2006) is used to formalize and verify problem frame instances. Alloy uses first order relational logic. Moreover, they use a technique called *requirements progression* to mechanically derive a specification from a given system-wide requirement and its associated domain assumptions. Here this would correspond to deriving control constraints from the system-level constraints and associated domain assumptions. In section 4.4, we use the notion of an *anti-system-level constraint*. This is similar to the notion of an *anti-requirement* introduced in (Lin, et al., 2003) to represent the intensions of a malicious user.

Combining formal reasoning and simulation is explored in (Gelman, et al., 2014), where the SAL model checker is used to guide the use of agent-based simulation via WMC (Sierhuis & Clancey, 2002). Specifically, the SAL model checker was used to model a known *automation surprise* associated with the Airbus A320 autopilot, which was then subsequently explored via WMC. They found that WMC provided psychological plausibility to the counterexample generated by SAL. Moreover, WMC provided multiple refinements to the given SAL counterexample. With the growth in autonomous systems that rely upon human interaction, e.g., self-driving cars, then tools that assist in the early identification of automation surprises will have an important role to play within the assurance process.

The use of formal methods in conjunction with STPA is not new. In (Howard, et al., 2019) (Colley & Butler, 2013) a methodology for incorporating the first three steps of STPA within the Event-B (Abrial, 2010) formal modelling tool is described. What distinguishes our proposal is the focus on identifying loss scenarios. Specifically, the use of formal modelling to identify abstract loss scenarios which are then elaborated via simulation.

4 Adding Formal Modelling and Simulation to STPA

We use an automatic car door locking example to illustrate our proposal. The example is inspired by Michael Jackson's aircraft braking system example[1] (Jackson, 1995). In our example, we are concerned with ensuring that while a car is moving its doors should be automatically locked and while the car is stationary, the doors should be automatically unlocked[2].

The task is to construct a Door Lock Control Unit (DLCU) that ensures the locking and unlocking of the doors occurs at the right time. A fundamental design assumption will be that when the car is moving the wheels will be turning. From the perspective of STPA, the controlled process involves the locking mechanism for the car's doors and the car's wheel sensors. The DLCU receives feedback in the form of *wheel-pulse* signals from the sensors when the wheels are turning. In terms of control, the DLCU will be able to send *lockDoors* and *unlockDoors* signals to the doors.

Below we follow the first three steps of the STPA hazard analysis technique as described above. We use the following propositions to bridge between the informal and formal:

WT: true when car *wheels are turning*, otherwise false.
DL: true when car *doors are locked*, otherwise false.
WP: true when there exists a *wheel-pulse signal*, otherwise false.
DS: true when there exists a *door-lock signal*, otherwise false.
MV: true when the car is *moving*, otherwise false.
MV_{pm}: true when the process model indicates the car is moving, otherwise false.

More realistically, these definitions would be developed hand-in-hand with the application of STPA.

---

[1] Jackson's example was related to an accident at Warsaw's Okecie International Airport in 1993 (Ladkin, 1994).

[2] Clearly there are flaws with this approach. For example, if a car was heading to the edge of a cliff with insufficient braking distance, then exiting the moving car would be preferable to driving over the cliff. However, the purpose of the example is to illustrate our proposal.

4.1 Define purpose and scope of the analysis (step 1)

From the perspective of people traveling in a car, clearly the loss of their lives or injuries due to safety failings would be personally catastrophic/devastating. In addition, the manufacturer would experience loss through the legal consequences of such safety failings, as well as potentially suffering reputational loss. For the purposes of the example, we will focus solely on the human perspective, i.e.

L-1: loss of life or injury.

We have associated the following two hazards to **L-1**:

H-1: a passenger opens their door while the car is moving.
H-2: a passenger is unable to open their door while the car is not moving.

H-1 may result is a person falling from a moving car leading to loss of life or injury, while **H-2** may prevent a person from leaving a stationary car when their safety is in jeopardy, e.g., the car is on fire. Following STPA, each hazard should be associated with a system-level constraint, i.e.

SC-1: if the car is moving then the doors must be locked (MV $\Rightarrow$ DL). [**H-1**]
SC-2: if the car is not moving then the doors must be unlocked. ($\neg$MV $\Rightarrow$ $\neg$DL) [**H-2**]

Note that composing the above two system-level constraints gives:

SC-3: the car is moving if and only if the doors are locked (MV $\Leftrightarrow$ DL). [**H-1, H-2**]

Note that we will use **SC-3** for the purposes of verification and the identification of loss scenarios.

4.2 Model the control structure (step 2)

As noted in section 2, we use the notation of a problem frame rather than a control structure in order to bridge between the informal and the formal. Figure 2 provides a problem frame for DLCU. Note that the problem frame makes explicit the domain assumptions and shared phenomena that logically link the system-level constraint with the control constraint. Note in particular that the design assumption that the movement of the car is equivalent to its wheels turning is explicitly represented.

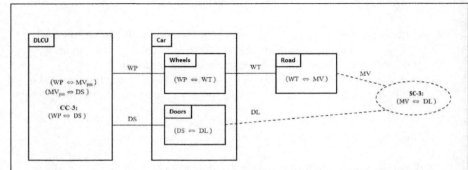

The above problem frame contains 3 domains, i.e., DLCU, Car and Road, where the Car domain has 2 subdomains, Wheels and Doors. The Road domain captures a key design assumption, i.e. (WT ⇔ MV). Domains interact via shared phenomena. For example, DLCU and Wheels share wheel-pulse signals (WP). The DLCU is a special domain; it represents the software controller, where its intended behaviour is specified via the control constraint **CC-3**. The dashed ellipse denotes the required system-level constraint, i.e., **SC-3**. The dashed lines indicate that the system-level constraint relates to the Road and Doors domains.

Fig. 2. A Problem Frame for DLCU.

In terms of responsibilities, DLCU has two; each represents a refinement of a system-level constraint, i.e.

R-1: enable door locks when the car is moving. [**SC-1**]
R-2: disable door locks when the car is not moving. [**SC-2**]

As described in section 2, each responsibility is associated with elements of a controller's process model and control logic. These associations for DLCU are given below. They include both the STPA informal descriptions as well as a logical counterpart:

Responsibility R-1:
Process model: car is moving (MV_{pm})
Control logic:
- if wheel-pulse signal then process model indicates car is moving ($WP \Rightarrow MV_{pm}$).
- if process model indicates car is moving then lockDoors control action is applied ($MV_{pm} \Rightarrow DS$).

Feedback: wheel-pulse signal (WP).

Responsibility R-2:
Process model: car is not moving ($\neg MV_{pm}$)

Control logic:
- if no wheel-pulse signal then process model indicates car is not moving ($\neg WP \Rightarrow \neg MV_{pm}$).
- if process model indicates car is not moving then unlockDoors control action is applied ($\neg MV_{pm} \Rightarrow \neg DS$).

Feedback: no wheel-pulse signal ($\neg WP$).

The final part of step 2 involves defining the control actions that are intended to achieve a controller's responsibilities. In the case of DLCU there are two:

CA-1: lockDoors (when door-lock signal then DS is true). [**R-1**]
CA-2: unlockDoors (when no door-lock signal $\neg DS$ is true). [**R-2**]

4.3 Identifying unsafe control actions (step 3)

A STPA style hazard analysis for the two control actions associated with DLCU is shown in Table 1. From the hazard analysis control constraints can be derived. The hazard analysis gives rise to the following two control constraints:

CC-1: if wheel-pulse feedback then door-lock signal (WP $\Rightarrow$ DS) [**CA-1**]
CC-2: if no wheel-pulse feedback then no door-lock signal ($\neg WP \Rightarrow \neg DS$) [**CA-2**]

Table 1. Identifying Hazardous System Behaviour for DLCU.

Control action	Not applied causes a hazard	Applied causes a hazard	Applied too early, too late or out of order causes a hazard	Stopped too soon or applied too long causes a hazard
lockDoors (**CA-1**)	Doors not locked when car moving (**H-1**)	Doors locked when car stationary (**H-2**)	Doors locked too late (**H-1**)	N/A
unlockDoors (**CA-2**)	Doors locked when car stationary (**H-2**)	Doors unlocked when car moving (**H-1**)	Doors unlocked too early (**H-1**)	N/A

Note that composing the above two control constraints gives:

CC-3: wheel-pulse feedback if and only if door-lock signal (WP $\Leftrightarrow$ DS) [**CA-1, CA-2**]

The STPA Handbook emphasizes the need for verification, i.e.

*'STPA generates the safety requirements (e.g., **SC-3**) and constraints for the automated control algorithm (e.g., **CC-3**). These must, of course, be verified to be correct.'*

As mentioned above in section 2, problem frames allow us to informally and formally verify the consistency between control constraints and system-level constraints. With regards to the consistency of **CC-3** and **SC-3**, Figure 3 provides an informal argument diagram, as well as a sequent style proof tree representation and our ELP representation.

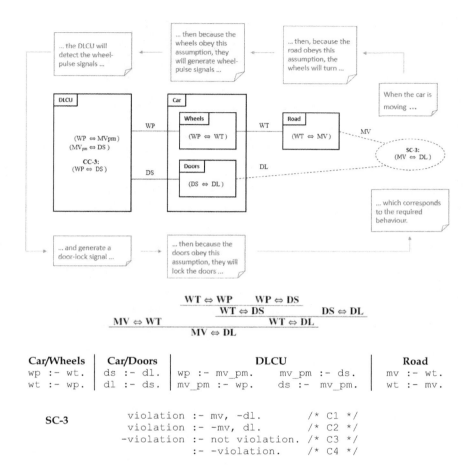

Clauses C1 and C2 define how **SC-3** can be violated. Note that the satisfaction of -dl and -mv requires explicit evidence for the negated propositions, i.e. failure to satisfy dl and mv is not sufficient to conclude that their negations hold. Clause C3 defines an explicit violation in terms of the failure satisfy violation. Clause C4 encodes the top-level proof: explicit evidence that there are no violation implies **SC-3** it true. The Extended Logic Programming demos are available at (Web-resource-ELP, 2022).

Fig. 3. Proving consistency between **CC-3** and **SC-3**.

4.4 Identifying loss scenarios (step 4)

As mentioned in section 2, we focus here on how loss scenarios could occur as a result of *flawed assumptions*. Specifically, we will focus on assumptions relating the environment in which a controller (i.e., DLCU) operates. Recall that a loss scenario results in the violation of a system-level constraint. In the case of our car example, the following proposition is equivalent to the violation of system-level constraint **SC-3**:

$$(MV \land \neg DL) \lor (\neg MV \land DL) \tag{1}$$

Note that the left disjunct of (1) relates to **H-1**, a passenger opens their door while the car is moving, while the right disjunction relates to **H-2**, a passenger is unable to open their door while the car in stationary. Here we focus on the left disjunct, i.e.

$$(MV \land \neg DL) \tag{2}$$

The verification proof given in Figure 3 relies upon three assumptions. The violation of each assumption gives rise to a distinct loss scenario that satisfies (2). Two of the scenarios correspond to component accidents, i.e., a wheel-pulse signal failure and a door-lock signal failure. Below we focus on the third assumption, i.e.

$$(WT \Leftrightarrow MV) \tag{3}$$

Specifically, we consider the case where (3) is replaced by:

$$(WT \Rightarrow MV) \land \neg(MV \Rightarrow WT) \tag{4}$$

This violation of (3) gives rise to a loss scenario where the car wheels are not turning but the car is moving, i.e.

$$(\neg WT \land MV) \tag{5}$$

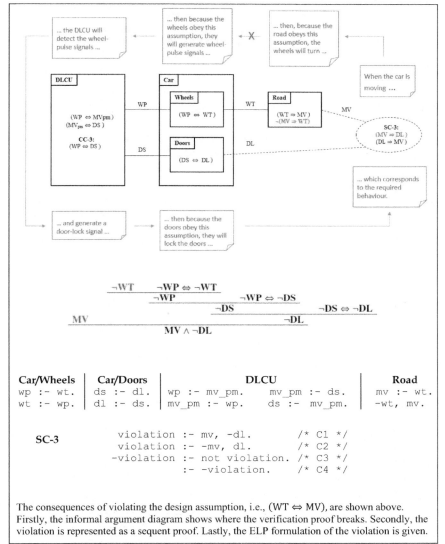

Fig. 4. A violation of the consistency between **CC-3** and **SC-3**.

As shown in Figure 4 replacing assumption (3) by (4) (within the **Road** domain) breaks the verification proof given in Figure 3. In Figure 5, we present an alternative formal model of the loss scenario. We replace **SC-3** by an *anti-system-level constraint*, i.e., **anti-SC-3**. While a system-level constraint specifies, 'something bad will never happen', an anti-system-level constraint specifies, 'something bad will happen'. The idea is that a scenario that achieves an anti-system-level constraint will denote a loss scenario.

So formal modelling provides a mechanism for generating abstract loss scenarios. However, it provides no insight into the feasibility of the generated loss scenarios. We use simulation to provide such insight. Simulators are rich in terms of their knowledge about real-world phenomenon. What we propose is to leverage this knowledge in order to elaborate an abstract loss scenario. In so doing, we are exploiting synergies that exist between formal modelling and simulation, and providing engineers with more realistic and accessible loss scenarios to consider.

In our car example, (5) provides the starting point for a simulation. Bridging the gap between the formal modelling (i.e., ELP) and simulation (i.e., CARLA) requires mapping the propositions MV and WT onto the corresponding parameters within the CARLA simulation space. CARLA provides various mechanisms to assist a developer to define a simulation. In terms simulation parameters, these are classified as follows:

- **Physics of a vehicle:** includes concepts such as torque, drag and friction.
- **Traffic behaviour:** predefined behaviours are provided, i.e. cautious, normal, aggressive, but a developer is free to define the behaviour of the vehicles within their simulation.
- **Perception:** weather conditions and the attributes of sensors, e.g. accuracy in poor lighting.

Given that the simulation space is significantly richer than the formal model, additional guidance is required in order to constrain CARLA. That is, we need access to domain knowledge about how stuff fits together as well as how stuff behaves in accordance to the laws of physics. For example, in our case study the concept of friction plays a key role in the loss scenario, i.e. wheel rotation requires friction between the car's tyres and the road surface. Such knowledge could be represented in terms of an *ontology*[3]. From the perspective of the case study, our use of CARLA is summarised in Figure 6. Note that the simulation demo and setup is available via (Web-resource-CARLA, 2022), which includes animations generated by CARLA. In Figure 7, a snapshot of the CARLA simulation development environment is provided.

---

[3] An ontology provides a means of representing and reasoning about domain knowledge.

Anticipating Accidents through Reasoned Simulation   261

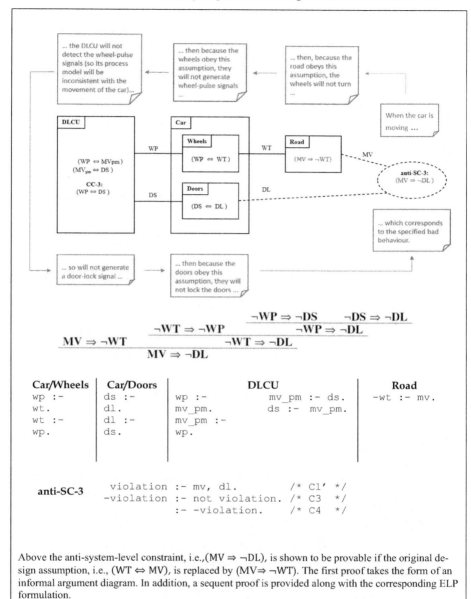

Fig. 5. Proving consistency between **CC-3** and **anti-SC-3**.

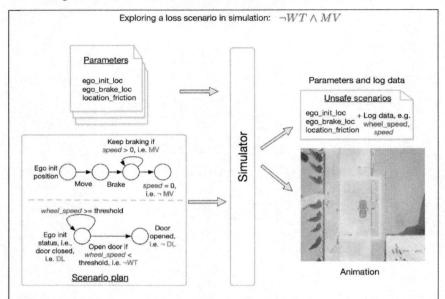

The inputs above on the left include, i) setup parameters for the simulation, and ii) template scenarios that encode the abstract loss scenario identified by the formal modelling. Note that while a graphical notation is shown, the developer uses a Python API to specify the inputs. The output from the simulator takes the form of i) an animation of the loss scenario, and ii) a log of all the relevant parameters. Note that a demo of the case study simulation is available via (Web-resource-CARLA, 2022), which includes the animations generated by CARLA.

Fig. 6. Exploring loss scenarios using simulation.

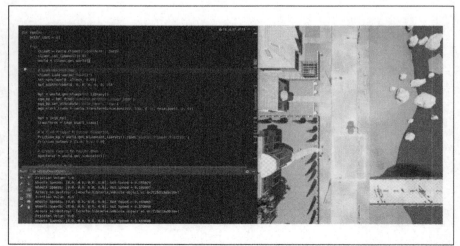

Fig. 7. A user's view of the CARLA simulator.

5 Future work

Two significant gaps need to be addressed in our future work. In order for our proposed approach to be effective and accessible, assistance is needed in bridging between STPA and ELP as well as ELP and CARLA. Tool support will be essential in managing the details. Full automation is not feasible or desirable. We envisage a level of dialogue between the tool and the user in order to ensure that each step in the translation is valid. In general, we believe that *Interactive Task Learning* (Appelgren & Lascarides, 2020) (Laird, et al., 2017) provides a promising approach to bridging these gaps. In addition to tool support, we also need to explore more realistic cases studies.

6 Conclusion

STPA is a popular technique that aims to assist engineers in analysing a system's control actions in relation to system-level hazards. The technique is structured and systematic but relatively informal compared to a formal method. Problem frames, if formalized, provides a mechanism for strengthening certain key informal aspects of STPA, i.e., verifying the consistency between system-level constraints and control constraints that are derived informally via STPA. Moreover, given such a formal verification we suggest that it can be used to identify loss scenarios related to invalid domain assumptions. Ultimately, engineers must be responsible for any safety related decisions. What we are proposing is an approach that assists engineers by providing them with simulated scenarios to review that are more realistic and accessible.

Acknowledgments The research reported in this paper is supported by the UKRI *Trustworthy Autonomous Systems Node in Governance and Regulation* (EPSRC grant EP/V026607/1).

References

Abrial, J.-R. (2010). *Modeling in Event-B: system and software engineering*. Cambridge University Press.

Appelgren, M., & Lascarides, A. (2020). Interactive task learning via embodied corrective feedback. *Autonomous Agents and Multi-Agent Systems, 34*, 1–45.

Colley, J., & Butler, M. (2013). A Formal, Systematic Approach to STPA using Event-B Refinement and Proof. *In proceedings of the 21th Safety Critical System Symposium.*

Dosovitskiy, A., Ros, G., Codevilla, F., Lopez, A., & Koltun, V. (2017). CARLA: An open urban driving simulator. *Conference on robot learning*, (pp. 1–16).

Gelfond, M., & Lifschitz, V. (1991). Classical negation in logic programs and disjunctive databases. *New generation computing, 9*, 365–385.

Gelman, G., Feigh, K. M., & Rushby, J. M. (2014). Example of a Complementary Use of Model Checking and Human Performance Simulation. *IEEE Trans. Hum. Mach. Syst, 44*, 576–590.

Howard, G., Butler, M. J., Colley, J., & Sassone, V. (2019). A methodology for assuring the safety and security of critical infrastructure based on STPA and Event-B. *Int. J. Crit. Comput. Based Syst, 9*, 56–75.

Jackson, D. (2006). *Software Abstractions: Logic, Language, and Analysis.* Cambridge, Mass. The MIT Press.

Jackson, M. (1995). *Software Requirements and Specification: a lexicon of practice, principles and prejudices.* Addison-Wesley.

Jackson, M. (2001). *Problem frames: analysing and structuring software development problems.* Addison-Wesley.

Ladkin, P. B. (1994). Report on the Accident to Airbus A320-211 Aircraft in Warsaw. *Report on the Accident to Airbus A320-211 Aircraft in Warsaw.*

Laird, J. E., Gluck, K., Anderson, J., Forbus, K. D., Jenkins, O. C., Lebiere, C., Salvucci, D., Scheutz, M., Thomaz, A., Trafton, G., Wray, R.E., Mohan, S. Kirk, J.R. (2017). Interactive task learning. *IEEE Intelligent Systems, 32*, 6–21.

Leveson, N. G. (2011). *Enineering a Safer World.* MIT.

Leveson, N. G., & Thomas, J. P. (2018). *STPA Handbook.* http://psas.scripts.mit.edu/home/materials/.

Lin, L., Nuseibeh, B., Ince, D. C., Jackson, M., & Moffett, J. D. (2003). Introducing Abuse Frames for Analysing Security Requirements. *RE* (pp. 371–372). IEEE Computer Society.

Seater, R., Jackson, D., & Gheyi, R. (2007). Requirement progression in problem frames: deriving specifications from requirements. *Requir. Eng, 12*, 77–102.

Sierhuis, M., & Clancey, W. J. (2002). Modeling and Simulating Work Practice: A Method for Work Systems Design. *IEEE Intelligent Systems, 17*, 32–41.

Web-resource-CARLA. (2022, Nov 1). *SSS'23 REASIM - CARLA Simulations.* Retrieved 11 1, 2022, from https://github.com/craigiedon/REASIM-Carla

Web-resource-ELP. (2022, Nov 1). *SSS'23: Extended Logic Programming (ELP) - Demo.* Retrieved 11 1, 2022, from https://colab.research.google.com/drive/1rkjQwKKYLDBooo5u5BWpmQtpK2IJB3v6?usp=sharing

Making the Water Visible: A methodology for exploring Systemic Change

Gill Kernick

Arup University, London, UK

Abstract *This paper articulates a methodology for exploring systemic change: Making the Water Visible. Developed as the author attempted to make sense of the Grenfell Tower Fire in which 72 people lost their lives and used as a structure for her 2021 book 'Catastrophe and Systemic Change: Learning from the Grenfell Tower Fire and Other Disasters'. The approach is codified in the hopes that others will further develop it to explore complex challenges and help us learn. It is as much a story of a complex journey through despair, grief, and sense-making as a method for making the water visible.*

1 Introduction

On the afternoon of the 14th of June 2017, more than 12 hours after the fire began, I watched my former home, Grenfell Tower (Figure 1), burn. Startling intense flames still exploding out of random charred windows.

I turned to Matthew Price, reporting on the fire for the BBC, and said 'I will do whatever it takes to make sure we learn, to – in some way - make those lost lives count.' That moment is as vivid for me as when I first saw the fire.

Naively I had imagined this desire to learn, would be shared and relentlessly driven across the government, housing, and built environment sectors. That the worst residential fire in London, a fire that killed seventy-two people in the UK's richest borough, would lead to swift, dramatic, and lasting change.

I was wrong.

Rather, I entered a labyrinth of despair.

From discovering the multiple failures to respond to the known dangers of external façade fires, to realizing that far from an isolated unsafe building, Grenfell was emblematic of decades of poor construction practices.

That thousands of families across the country, were living in unsafe buildings. And neither the government, housing sector nor industry were going to provide

© Gill Kernick 2023.
Published by the Safety-Critical Systems Club. All Rights Reserved.

the bold and courageous leadership needed to resource rapid remediation and ensure people could feel safe in their homes (for e.g., House of Commons 2020).

Fig. 1. Grenfell Tower 2012. I lived on the 21st floor (from 2011 to 2014, before the refurbishment that added a flammable façade (Wright 2012)

I had to confront, that, despite spin and rhetoric to the contrary, history predicted that we would not learn. That Grenfell, like so many tragedies before it, would not lead to meaningful change.

Predictably, we would conduct an Inquiry, identify key lessons, and fail to fully implement them or transfer them to other contexts. Yes, we would make some piecemeal changes, but the underlying issues such as failures to heed warnings, listen to those at the sharp end or deal with imbalances in power would not shift. These systemic issues were not limited to Grenfell, but to our inability to learn from disasters more broadly.

I realized that to fulfil my promise, rather than ask what could we learn from Grenfell, I had to ask a different question –Why does our failure to learn make sense? Until we understood this, we would never prevent the devastating human suffering whether caused by a pandemic, a plane crash, a flood, or a tower block fire.

This paper explores the process I adopted to help me answer this question. Beginning by exploring two issues that are at the heart of our failure to learn, and then offering a framework for exploring complex change by making the water visible. The concluding section explores the unexpected places where I have found hope.

As a warning, this is a retrospective articulation that bears little resemblance to the messy emergent process that took place over a three-year period. Making the water visible is never going to be a neat and tidy methodology. I found myself immersed in a world that at times was emotionally and intellectually overwhelming. I had to learn to sit with the mess; to allow things to emerge, to dwell in 'not knowing', and to start over – again and again and again. Until some new thread or thought helped me make sense of things in new ways and changed me in the process.

2 The Heart of the Matter

After years of research and reflection, my conclusion is that two interconnected issues lie at the heart of our inability to learn: Our failure to understand the nature of complexity, and an over-reliance on piecemeal versus systemic change.

2.1 The nature of complexity

Frameworks such as VUCA (volatile, uncertain, complex, and ambiguous) environments (Bennett and James Lemoine G, 2014), wicked (seemingly unsolvable) problems and complex adaptive systems are all indicative of a growing understanding of how to operate effectively in an increasingly complex world.

Many of our traditional, top-down, bureaucratic, and mechanistic ways of thinking, that are grounded in mythical cause-and-effect narratives such as 'we'll roll out this regulation and it will change behaviour', are becoming redundant.

And yet, our response to disasters largely comes from this paradigm. For example, in the UK, our pandemic response focussed on top-down regulations but failed to consider the impact on trust and behaviours of ministers and advisors violating lock down rules (Fancourt et al. 2020).

In complex systems, change is emergent, and can't be directed or controlled in a linear predictable fashion. Outcomes are unpredictable and we can't always retrospectively understand cause. Interactions are nonlinear and minor changes can produce major consequences. Additionally, elements in a system adapt and co-evolve based on their interactions with one another and with the environment (for e.g., Chan 2001; Snowden & Boone, 2007).

These characteristics are highlighted by events such the creation of a global youth movement for climate change stemming from one young girl's protest outside the Swedish parliament (Britannica, 2022). And in how local communities created support networks in response to Covid often far more rapidly and effectively than government interventions.

If we're serious about preventing disasters, we must improve our ability to operate effectively with increasing complexity. Skills such as complex problem solving, collaboration, critical thinking, creativity, emotional intelligence, and cognitive flexibility will become more important (Whiting, 2020).

This failure to understand and operate effectively in an increasingly complex world, is closely connected to overly focussing on piecemeal versus systemic change - the second factor at the heart of our inability to effectively learn from, and therefore prevent, disasters.

2.2 Piecemeal versus systemic change

Most of our responses to disasters are piecemeal: changing parts of the system but not the system itself. Piecemeal change will happen post-Grenfell, and it is critical that it does. I am not at all confident, though, that we will see systemic change.

Systemic change requires shifting the conditions that are holding the problem in place. Such problems 'are entrenched and perpetuated by the status quo of power, institutional culture, social expectations, myth and narrative' (Draimin and Spitz, 2014).

Rather than asking what is wrong with the system (which gives you piecemeal answers), this is best done by considering that the system is functioning perfectly and to observe and discover what it is perfectly designed to produce. Table 1 highlights some key differences between piecemeal and systemic change.

Table 1. Piecemeal versus Systemic Change

	Piecemeal Change	**Systemic Change**
Intent	Solving a piecemeal issue	Shifting the conditions holding the status quo in place
Question	What is wrong with the system?	What is the system perfectly designed for?
Assumption	Controllable, predictable world	Complex, emergent world

	Piecemeal Change	**Systemic Change**
Access to change	Fix what is wrong	Make the Water Visible: grapple with the messy kaleidoscope
Approach to change	Technical Solutions (If I do 'x', 'y' will happen)	Disrupting the status quo, experimenting (If I do 'y', what will happen?)
Leadership Style	Bureaucratic, command and control, rules based	Organic, emergent, values and principles based
Requires	Traditional Expertise	All Stakeholders tacit expertise

By way of example, let us consider the issue of building materials, which we know contributed to the fire. A piecemeal approach would attempt to control what specific materials were used on high-rise buildings which are important and critical changes. But they are not sufficient.

A systemic approach to the issue of building materials would require grappling with some messy issues. These include the role of political lobbying and self-funding testing and certification bodies; the trade-offs and decisions being made in the interests of sustainability; and the limitations of siloed governance and regulations.

This would reveal far more complex challenges than banning certain building materials or tightening up construction product testing and classification. It would require a rethinking of which behaviours we consider acceptable and which we do not.

Piecemeal change is important, and it drives incremental improvements, it is also relatively easy: you identify what went wrong and then put plans in place to correct it. Solutions tend to be technical in nature and appropriately rely on traditional expertise.

To explore systemic change, we need a different approach.

3 Making the Water Visible

To help understand why our failure to learn from disasters makes sense I developed a methodology for exploring systemic change that is codified here, so that others can use and build on it. Developed to consider disasters, it could be applied to any complex issue.

The fish parable provides a good analogy to the challenges of exploring complex issues.

> Two young fish are swimming along, when an older fish swims past and asks them 'How's the water?' The two young fish continue swimming for a while, and then one looks at the other and says, 'What's water?'

The first step in systemic change is to reveal the conditions holding the problems in place: to make the water visible, illuminating the systemic forces at play and grappling with this 'messy kaleidoscope' (Kania et al, 2018). This grappling, in and of itself, is transformative as new ways of thinking and viewing the world present themselves as trails of thoughts to be followed and explored and developed.

There are four components to making the water visible: the principles, the approach, the framework, and the questions.

3.1 The principles

The following principles underpin the methodology.

- Traditional bureaucratic, linear cause and effect ways of thinking and leading are ineffective in enabling systemic change.
- Piecemeal solutions don't cause systemic change and can have unintended consequences.
- The system is functioning perfectly to produce what it is currently producing; we need to make it visible to reveal the messy kaleidoscope of factors at play. For example, the penguin pool example below reveals issues Incremental Thinking, where multiple small changes are made, each not a problem in itself, but the accumulation of changes causes a major issue. And the consequences associated with the loss of original vision and design intent.
- Systemic change requires disrupting the status quo and given the unpredictable nature of emergent systems this requires an experimental or 'seed planting' approach.

Fig. 2. The Penguin Pool at the London Zoo, designed by Berthold Lubetkin and engineered by Ove Arup. © Arup

As an example. The Penguin Pool at London Zoo (Figure 2) completed in 1934, is an icon of modern British architecture and engineering. But, in 2004, the penguins were relocated due to getting bumblefoot, from micro abrasions caused by walking on concrete. In a fascinating letter, John Allan, the architect who restored the pool in the Eighties, explained that rather than issues with the original design, decisions had been made by the zoo that led to this outcome. The original largely rubber paving, designed for the penguin's comfort, was replaced with concrete. A layer of quartz was added to the ramp surfaces for the benefit of the keepers, but discomfort of the penguins. And the original birds for which the pool was designed preferred to huddle but were replaced by Humbolt penguins, which prefer to burrow, making the original nesting quarters unsuitable (The Reader, 2019).

This sweet illustration of the principles shows the limitations and unintended consequences of piecemeal solutions, the ineffectiveness of linear cause and effect ways of thinking and the need to step back and view issues holistically. It reminds me of building on floodplains and other piecemeal decisions that have led to untold devastation and suffering.

These principles give rise to the Approach, designed to move beyond these ways of thinking.

3.2 The approach: Enquiry and Sense-making

The approach is grounded in the art of enquiry and sense-making, with the intent to enable action and change. The use of stories and critical friends are essential.

'Enquiry' is used to differentiate it from the more formal investigation processes or 'Inquiries'. Authentic enquiry requires inhabiting the space of 'not knowing', in the words of Socrates: 'I know that I know nothing'.

Sense-making, a term coined by Karl Weick, is tangled with an enquiry-based approach. Sense-making is not an academic exercise and does not require your neutrality. Distinct from analysis, interpretation, or exploration it is a creative process that 'makes sense' of unknown and complex domains in a way that enables action (Ancona, 2012).

This intent to engender action and change is critical. The two overarching enquiries I sat with were:

- Why does our failure to learn make sense?
- What will it take to enable systemic change?

3.3 The Framework

The Grenfell Framework for Systemic Change' (Figure 3) provides a structure for making the water visible. A way of exploring the messy kaleidoscope of factors at play. It looks through four lenses: Governing and Operating frameworks, and Obvious and Obscure elements.

Governing frameworks provide the architecture for decisions and thinking. Operating frameworks guide how we function. The Obvious elements are the lenses we usually look through, such as regulations or investigations. Many analyses focus solely on these, which fails to create a holistic picture. The Obscure elements include issues such as relational tensions, power, narratives, culture, biases, and issues of trust.

These lenses sit behind four quadrants.

- The Foundational Quadrant considers what structures are in place to prevent catastrophic events.
- The Behavioural Quadrant considers what mechanisms are in place to prevent and respond to catastrophic events.
- The Relational Quadrant considers how relational issues contribute to catastrophic events and our ability to learn.
- The Contextual Quadrant considers the contextual aspects that impact our ability to prevent and learn from catastrophic events.

Making the Water Visible 273

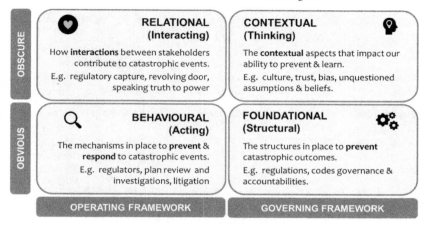

Fig. 3. The Grenfell Framework for Systemic Change

3.4 The Questions

Within each of these quadrants, four questions are considered, that are designed to give access to the messy kaleidoscope of factors at play, to facilitate sense-making and reveal new opening for acting, experimenting and seed planting.

What are the known issues? This entails more traditional research methodologies to expand understanding about issues that, despite being known about, have persisted over time. For example, complex delivery mechanisms in local and central government that prevent clarity of accountability. Sitting with the picture this presents helps explore the remaining questions.

What myths are contributing to our inability to learn? This is designed to reveal widely held (and mostly unquestioned) beliefs that hinder learning. Holding onto these myths justifies an over reliance on the obvious responses that fail to address systemic issues.

What are the conditions holding our inability to learn in place? Through a process of sense-making this enquiry begins to reveal the key factors holding conditions in place and inhibiting systemic change. Rather than working to shift these factors, which are entrenched, considering them as 'givens' and working to enable change despite them will likely be more effective.

How might we enable systemic change? Sitting with the messy kaleidoscope allows less obvious opportunities to disrupt the status quo to emerge. Observing what people are doing and the impact (or lack of impact) is a critical part of this. Given the complexity involved these disruptive opportunities will change and evolve over time and are best viewed as areas to explore and experiment with, rather that guaranteed solutions to change.

As an illustration, Table 2 summarises the findings that emerged through this process over a three-year period, regarding Grenfell (Kernick, 2021).

Table 2. Grenfell: Why does our failure to learn make sense?

Question	Foundational	Behavioural	Relational	Contextual
Known Issues What are some of the known issues that have persisted over time?	Challenges with legislation and regulation in the UK Competing objectives and tensions Political agendas and the role of lobbying	Challenges with government implementing effective preventative guidance. Weak or ineffective regulators. No mechanism for ensuring inquiry recommendations are implemented	Regulatory capture Groupthink Herd behaviour The challenges with speaking truth to power Issues with meaningful public consultation	Bias and decision making The competing tensions faced by decision makers The role of the media Deception and issues of trust
Myths What are the myths we hold onto that prevent us from looking at issues systemically?	That regulations guarantee safe outcomes.	That the default world is perfect and error-free, and mistakes are abnormal.	That the 'softer' relational issues aren't that important.	That you can enable systemic change without shifting deeply held assumptions and beliefs.
Why don't we learn? What is the condition holding the status quo in place	Our Persistent failure to address issues in governance and accountability	Our obsession with blame and blame avoidance	Our failure to effectively rebalance power	The lack of political intent and will to enable real systemic change
How could we enable systemic change? What is our biggest opportunity to disrupt the status quo?	Improve our capability to deal with complexity and ambiguity	Ensure fairly borne consequences, where those that enabled disasters (versus the victims) bear the consequences for remediation etc.	Tap diverse and distributed knowledge	Create safe spaces to engage with and challenge deeply held views

Throughout the process, I was surrounded by critical friends and mentors, which with hindsight was invaluable. They validated and challenged my sense-making and helped shape my thinking. They also gave me hope.

One of my mentors was Jim Wetherbee (Figure 4), a retired American Naval officer, test pilot and NASA astronaut. On a day when I felt particularly helpless, Jim said: 'The doors of the people who should be talking to you will likely be closed, stop knocking on them. Look for the open doors.' This took me out of linear thinking, into the emergent maze of change and complexity. I've learned that systemic change requires disruption – that I need to shift how I think and act. I've come to see that kindness can be more disruptive than aggression, that compassion can be more impactful than taking positions. I've learned to plant seeds, and not to worry about which ones live and which ones die.

When I say the intent of the approach is to enable change – it is this kind of change, change that will disrupt the status quo, that is about planting seeds and experimenting and discovering which seeds grow and which don't.

Fig. 4. Dr. Jim Wetherbee (Photo Credit: NASA)

This approach to making the water visible is existential in nature – you will be changed, your beliefs and biases will (and should) be challenged and exposed and questioned and altered.

Now to hope.

4 Of hope

I want to end in the same place I started. With Grenfell. This time with the last day of the Inquiry's Evidentiary Hearings in July 2022 This final module sets out the individual circumstances surrounding each death. I am here to attend the hearings of my former neighbours, the El Wahabi's (Grenfell Inquiry, 2022). They all died in their home, flat 182, on the 21st Floor. Abdulaziz's sister Hanan, lived on the 9th floor and escaped from the tower with her husband and two children.

The conditions on floor 21 were rapidly deteriorating and changing. It is against that backdrop that we hear about the numerous emergency calls that were made, and the advice given to stay in their flat and wait for the fire service.

One call to a Control Room Operator lasted for 59 minutes. Beginning at 1:38 (less than an hour after the fire started). The operator repeatedly tells the family to stay in their flat saying that the firefighters would be coming soon and would have oxygen. As the conditions worsen, she tells them to close windows, cover their faces and move to the bedroom.

Just after 2:21, the control room operator was told by the family that the fire was in the corridor of the flat and that the smoke was coming into the bedroom. She told the family to cover their mouths and to get as close to the floor as they could.

In another call at 2:47 the family say they are now under the bed. Nur Huda says, "We are dying, and we can't get out". They are told the fire service is on its way. Abdulaziz says "I could have got out a long time ago, we could have but they said stay in the flat, stay in the flat. We stayed in the flat; we didn't leave."

In the final calls with the control room the family are advised to leave but they say it is too late as there is too much smoke and they cannot breathe.

According to the archaeological records, they all died in one of the bedrooms. They were lying close together.

Juxtaposed with these calls, outside the tower, Hanan and other members of her family were having desperate phone calls with the El Wahabi's, pleading with them to 'get out'. They stood outside the tower, watching the horror of the spreading flames unfold, they could see Abdulaziz and his family inside their flat at the windows.

According to protocols, Flat 182 should have been a priority for evacuation as it was known that there were children present. There were no firefighter crews deployed or making it to that floor.

In the words of counsel Mary Monroe:

> One branch of the family survived: Hanan, her husband, her children. They did not make any emergency calls, they evacuated and survived. They live to mourn and question why.

I struggle to reconcile the faith the El Wahabi's placed in the control room operator's advice. Listening to them over their own families. That afternoon has changed me in ways I do not yet understand.

It has made me think very differently about the burden of leadership. About the advice we given and the decisions we take. About the very real consequences of our failure to learn from tragic events.

My journey post Grenfell has been one of despair and hope. I adored living there, my memories are of the sounds of children playing in the lift lobby and the unimaginably exquisite sunsets. I have been horrified and left despairing at the racist attacks on the community.

The worst being the emergence of a YouTube video of a group burning an effigy of Grenfell just over a year after the fire. I struggle to understand the context surrounding this as I imagine people taking the time to make the model, add a label - Grenfell Tower. And then laugh and joke as they watch it burn.

I have despaired at the lack of trustworthiness of institutions and governments and organisations as the gut-wrenching revelations of the inquiry have unfolded over the last years.

I have despaired at the how risk is born so unequally and how those most vulnerable are most exposed and increasingly so as the impacts of climate change are so apocalyptically being felt.

At times it has been hard to move beyond despair, but I have found hope in unexpected places, two of which I will share: Grief and the Democratisation of change.

4.1 Grief

We talk too little about the transformative power of grief. Of the power and resilience and determination of those most impacted by disasters. Of the ability of those who lost their homes and loved one to drive change.

But also of the grief, of those, like me and many reading this, who are on the margins of disaster and whose souls are touched in ways that compel action and change. Grief that gives courage and wisdom and an unrelenting drive for progress and learning.

Recent research (Crosweller, 2022) found that disaster managers felt that loss and suffering were not well understood.

The same research found that the leadership traits most needed after disasters include compassion and kindness and courage and wisdom (Figure 5). I have found hope in many interactions, many with people I have never met that have demonstrated these leadership traits and touched my heart forever.

Fig. 5. Characteristics that constitute the virtues of disaster management leadership (Crosweller, 2022).

I wonder if there is more that we could do to build these characteristics and develop our capacity for a new way of leading.

Edward Daffarn, the Grenfell survivor who campaigned for change prior to the fire and was labelled a rebel resident, said in his evidence at the Grenfell Inquiry (Grenfell Inquiry, 2021):

> They didn't treat us with respect or humanity or empathy, and if they had done, we wouldn't be sitting here now.

I would encourage us to talk more openly about grief, understand its transformative power and develop our ability for a new way of leading.

4.2 The democratisation of change

There is immense hope in what I call the democratisation of change. An increasingly complex world and the democratising effect of social media has given us all the opportunity (and perhaps duty) to contribute significantly to learning and preventing disasters.

I no longer believe it is the sole remit of governments to drive change, we have a responsibility too. If you want to change things then you need to let a thousand flowers bloom, some will thrive, some will not; you cannot determine in advance what will work…

As complexity expert Dave Snowden (2020) says:

> You want a wildflower meadow, not a formal garden

We can all be disruptive seed planter and gardeners – in whatever way we can – and tiny-step-by-tiny-step change will happen.

5 Concluding Remarks

My intent has been to articulate the 'making the water' methodology so that others may use and develop it further.

At the heart of our failure to effectively learn from, and therefore prevent disasters, lies a failure to embrace the interconnected worlds of complexity, and systemic change. Embracing these worlds demands a radically different style of leadership, one grounded in principles and values and that engenders organic and emergent thinking and solutions.

The Making the Water Visible methodology offers a framework for enquiry and sense-making that might help us move towards goodness and meaningful change. To a world where a resident's 'lived expertise' is valued as greatly as the technical advice of an engineer. Where experimentation, innovation and disruption are the norm. Where not-knowing, humility and curiosity are valued over control, bureaucracy, and authority.

Where hope lives in fellow seed-planters, seeking a world we can't yet see.

A world where the terrible human suffering of disasters, is in fact honoured through meaningful and lasting change.

My hope is that engaging with this kind of approach to make the water visible may help us, in the words of Rebecca Solnit (2019),

> remake the world, and ... do so mostly by the accretion of small gestures and statements and the embracing of new visions of what can be and should be.

But ultimately, as I sit with images of the archaeological remains of families dying together with the parents forming protective rings around their children, the effectiveness of our responses to disasters is bounded by the soundness of the decisions we make and the moral leadership and courage we bring to making them. As Counsel to the Inquiry Richard Millet said.

> ... every decision, every act, omission, interpretation, understanding, practice, policy, protocol, affects someone somewhere, someone who is unknown and unseen, but who is an adored child, a beloved sister, a respected uncle, a needed mother. (Grenfell Inquiry, 2022)

May those words ring in our ears and may the voices of those not here, help us.

References

Ancona R, 2012, Sensemaking: Framing and Acting in the Unknown, MIT Sloan School of Management, The Handbook for Teaching Leadership, Chapter 1, Sage Publications.

Bennett N and James Lemoine G, 2014, What VUCA Really Means for You, Harvard Business Review, January–February 2014

Britannica, The Editors of Encyclopaedia. "Greta Thunberg". *Encyclopedia Britannica*, 1 Jan. 2022, https://www.britannica.com/biography/Greta-Thunberg. Accessed 5 December 2022.

Chan S, 2001, Complex Adaptive Systems. Research Seminar in Engineering Systems, 31 October – 6 November 2001, MIT.

Crosweller M, 2022, Disaster management leadership and the need for virtue, mindfulness, and practical wisdom, Progress in Disaster Science, Volume 16, December 2022, published by Elsevier, https://www.sciencedirect.com/science/article/pii/S2590061722000357#f0005 Under https://creativecommons.org/licenses/by-nc-nd/4.0/

Draimin T and Spitz K, 2014, Building Ecosystems for Change: How do we collaborate to create ecosystems that support innovation for systems change? Social Innovation Generation and Oxfam.

Fancourt D, Steptoe A and Wright L, 2020, The Cummings effect: politics trust and behaviours during the Covid-19 pandemic. The Lancet, 6 August 2020. https://doi.org/10.1016/S0140-6736(20)31690-1

Grenfell Tower Inquiry Day 118 April 21, 2021, Edward Daffarn, p.242. lines 20-22. https://assets.grenfelltowerinquiry.org.uk/documents/transcript/Transcript 21 April 2021.pdf

Grenfell Inquiry, Day 308, July 21, 2022, El Wahabi hearings pp.89 – 120. https://assets.grenfelltowerinquiry.org.uk/documents/transcript/Transcript 21 July 2022_0.pdf

Grenfell Inquiry, 2022, Day 308, July 2022, QC Richard Millett, pp 122, 20-22. https://assets.grenfelltowerinquiry.org.uk/documents/transcript/Transcript 21 July 2022_0.pdf

Kania J, Kramer M, Senge P, 2018, The Waters of Systems Change, FSG, https://www.fsg.org/wp-content/uploads/2021/08/The-Water-of-Systems-Change_rc.pdf

Kernick, G, 2021, Catastrophe and Systemic Change: Learning from the Grenfell Tower Fire and Other Disasters, London Publishing Partnership, p 208

Snowden D, 2020, Self-destructive tendencies, Cognitive Edge, 3 September

Snowden D and Boone M, 2007, A Leaders Framework for Decision Making. Harvard Business Review.

Solnit R, 2019, Introduction in 'Whose Story is This? Old Conflicts, New Chapters, London, Haymarket Books.

The Reader, 2019, Penguin pool surely has not 'had its day'. Evening Standard, 7 January 2019. https://www.standard.co.uk/comment/letters/the-reader-penguin-pool-surely-has-not-had-its-day-a4032146.html

Whiting K, 2020 'These are the top 10 job skills of tomorrow', World Economic Forum, 21 October 2020. https://www.weforum.org/agenda/2020/10/top-10-work-skills-of-tomorrow-how-long-it-takes-to-learn-them/

Wright I, 2012, Grenfell 2012, CC BY-SA 2.0 https://creativecommons.org/licenses/by-sa/2.0 ,via Wikimedia Commons.

Machine safety conformance limitations for highly automated and autonomous heavy-duty mobile machinery

Aimée M.R. de Koning, Reza Ghabcheloo

Tampere University

Abstract *There is a great drive and incentive in industry to increase the level of automation in heavy-duty mobile machinery, but further progress is slowed down due to a lack of regulations and division of legal responsibilities, on top of the limitations of system capabilities in terms of reliability, maintainability, performance, and available technologies. In higher levels of automation, the operator is no longer in full control of the machine, and the machine itself becomes the controller. The newly emerging requirements for safety are not covered by existing standards leading to difficulties for manufacturers to embed a justifiable level of safety into their machinery. In this paper, we first provide a survey on relevant recent research efforts towards safer highly automated and autonomous systems. We then discuss the conformance process and emerging limitations of existing EU machine safety regulations in relation to an increase of automation in heavy-duty mobile machinery. Guided by a clarifying example we then identify six topics in existing EU machine safety regulations, limiting the conformance of machinery a) run-time failures, b) algorithmic failures, c) convoluted architectural design patterns, d) data-driven intended behaviour, e) quality integration and f) formal verification limitations. We assert that reaching future compliance of highly automation and autonomous heavy-duty mobile machinery is achieved through overcoming the aforementioned limitations.*

1 Introduction

The production of machinery represents a core element of the broader mechanical engineering sector in Europe's largest and most competitive industry (EU Manufacturers, 2022). For machinery to be marketed in the European Economic Area (EEA) the manufacturer ensures that their products comply with EU safety, health, and environmental protection requirements. It is their responsibility to

© Aimée de Koning and Reza Ghabcheloo, Tampere University 2023.
Published by the Safety-Critical Systems Club. All Rights Reserved.

carry out a conformity assessment, set up all the necessary technical documentation, issue the EU declaration of conformity, affix the CE marking on a product, and where necessary have conformance checked by a notified body (CE Marking for the machinery industry, 2022). The involvement of a notified body is only necessary for machinery and parts thereof mentioned in Annex IV of the Machinery Directive (EU commision; Machinery Directive, 2006), which includes for example the logic units ensuring safety functions. The procedure of assessing the conformity of machinery is described in Article 12 of the machinery directive, stating that if conformity is assessed through a notified body a system or parts thereof become certified, otherwise the manufacturer must ensure conformance through validation, by its own specialized safety conformance engineer(s) or an authorized representative (EU commision; Machinery Directive, 2006). When a safety-critical situation occurs during the operation of such a machine -and it evolves into a legal dispute to determine culpability- conformance with the Machinery Directive (EU commision; Machinery Directive, 2006) and a number of standards such as EN ISO 13849-1:2015 (13849-1:2015, 2015), EN 62061:2005 (62061:2005, 2005) serve as a guide to establish compliance with regulatory requirements and thereby determine if the manufacturer is liable for damages.

1.1 Machine safety conformance terminology

Machinery is defined as an assembly consisting of linked parts or components, at least one of which moves, and which are joined together for the execution of a function (EU commision; Machinery Directive, 2006). Figure 1 shows an example of heavy machinery for the purpose of excavating, i.e. an excavator. This image is taken at the mobile hydraulics laboratory of Tampere University in Finland which provides facilities for field robotics research and holds several instrumented heavy-duty mobile machines. The manufacturer is expected to design a functionally safe machine which covers all potential risks and hazards that can reasonably be expected to occur during a machine's lifecycle, including those stemming from potential misuse (EU Manufacturers, 2022). The concept of *functional safety* in machinery is best defined in EN 62061:2005 section 3.2.9 (62061:2005, 2005) as part of the safety of the machine and the machine control system which depends on the correct functioning of the safety related electrical control system, responsible for the reliable availability of safety functions. A *safety function* (SF) as defined in (12100:2010, 2010) is a function of a machine the failure of which can result in an immediate increase of risks and established through a defined set of safety related requirements that must be embedded into a machine's design. Here a *risk* is considered to be a combination of the probability of occurrence of harm and the severity of that harm (12100:2010, 2010). A *hazard* is considered the potential source of a harm -physical injury or damage to health - where a safety-critical situation expresses the circumstances in which a person is exposed to at least one hazard (12100:2010, 2010). The goal of a safety function is to appropriately mitigate and where possible avoid a safety-critical situation that could occur with the use of the machine, and in conventional methods of compliance must be traced back to a safety requirement.

These safety requirements of a machine emerge during a risk assessment and hazard analysis as guided by the relevant machine safety standards and are linked

Fig. 1. Heavy machine for the purpose of excavating, i.e. an excavator. (Renval, 2022)

to the implementation of a safety function. The increasing level of automation in road vehicles revealed limitations within the scope of the conventional definition of safety as defined in road vehicle standard ISO 26262:2011 (12100:2010, 2010), which defines safety as functional safety similar to the definition established in machine safety standard EN 62061:2005 (62061:2005, 2005). In the automotive domain this has led to expanding the scope of the definition of safety to cover safety of the intended function as is described in section 3.25 of road vehicle standard ISO 21448:2022 (21448:2022, 2022). *Safety of the intended function* (SOTIF) is defined as the absence of unreasonable risk due to hazards resulting from functional insufficiencies of the intended functionality or its implementation (21448:2022, 2022). SOTIF refers to safety-critical situations that can occur outside of the defined scope of safety functions e.g. a safety function which covers automated breaking capabilities which does not take road conditions into account. While the safety function remains available, its functional execution does not reduce the risk of a safety-critical situation as the calculated break path might be shorter or longer in reality depending on the road conditions. Therefore this safety function falls short of its intended functionality e.g. avoiding collision by safely reducing the speed of the system. To the best knowledge of the authors there is currently no machinery specific standard available that guides similar SOTIF related challenges. Therefore when discussing SOTIF, we will avail ourselves of the definition established in ISO 21448:2019.

1.2 European research efforts related to the safety of autonomous systems

Across Europe research efforts towards safe, highly automated, and autonomous machinery is an active subject in several research consortia. *The European research and training network for safer autonomous systems* or *SAS* is a consortium made up of academic institutes and companies in western Europe across all domains including machinery. Funded by the European Union's EU Framework Program for Research and Innovation, focusing on the questions for safety that arise when an autonomous system interacts with humans. Started by TTTech Auto in central Europe, *The Autonomous* is a consortium with the goal of establishing a reference for safety in autonomous road vehicles. They annually organize events with invited speakers hosted in Vienna. The consortium is mainly focused on the automotive domain and is supported by multiple automotive industrial partners establishing working groups to tackle specific challenges related to the reliability of such systems. The *MORE* consortium is another European research and training network on highly automated and autonomous heavy duty mobile machinery, comprised of world leading OEMs and technology providers of heavy-duty mobile machinery. Their research efforts are best described in (Machado et.al., 2021). What ties all these individual efforts together is the desire for trustworthy and safe autonomous and highly automated systems in conformance with EU Machine Safety legislative requirements.

1.3 Related work

The EU regulates safety and reliability across a myriad of industries through domain specific directives and the harmonisation of system specific standards with EU legislative requirements. Both automotive systems covered by specific Motor vehicle safety regulations and robotic systems which fall under the Machinery Directive are most closely related to that of heavy-duty mobile machines and face similar conformance limitations as the level of automation increases. Safety in the automotive domain is established through the European Commission's work on motor vehicle safety, dealing with the safety of vehicle occupants and vulnerable road users (EU Safety in the Automotive Sector, 2022). Automotive standard ISO 26262:2011 is wildly referenced in relation to the functional safety of automotive vehicles, as well as generic standard IEC 61508:2010 in relation to the conformance of functional safety of electrical/electronic/programmable electronic safety-related systems. Robotic systems fall under the scope of the machinery directive and share references to generic standards, but robotic system specific harmonised standards differ from those required for heavy-duty mobile machinery.

A key factor with road vehicle standard ISO 26262:2011 is that the human driver is ultimately responsible for safety, thus limiting the conformance of road vehicles with higher levels of automation. Authors (Philip Koopman and Michael Wagner, 2016) have identified six road vehicle specific conformance limitations alongside the safety development lifecycle in accordance with ISO 26262:2011, namely, i) sufficient coverage[1] in the presence of increased complexity of requirements, ii) traceability[2] of requirements, iii) safe integration with the rise of probabilistic and data-driven components, iv) formal verification, v) explainable justification towards a systems safety, and vi) the need for fail-operational systems. In (Philip Koopman and Uma Ferrell, 2019) the authors continue to discuss how part of these limitations with automotive standard ISO 26262:2011 are addressed by ISO 21448:2019 Road vehicles — Safety of the intended functionality. ISO 21448:2019 discusses safety in terms of mitigating risks associated with the emergence of undesired behaviour due to a lack of coverage[1] often brought on by limitations in perception systems. The authors evaluate ISO 21448:2019 discussing that this standard still leaves gaps in appropriately addressing legislative incompatibilities due to the use of probabilistic and data-driven components in higher level automated systems with respect to their testability and traceability of safety requirements. These gaps inspired their proposal of ANSI/UL 4600, a Standard for Safety for the Evaluation of Autonomous Products. It is built on the strengths of ISO 26262:2011 and ISO 21448:2019, in an attempt to bridge the gap of evaluating highly automated and autonomous systems through the use of so-called safety case arguments.

Limitations with the conformance of systems with a high level of automation and complexity also persist in the robotics domain. The authors (Aude Billard and Pericle Salvini, 2021) reviewed EN ISO 13482:2014 — Safety requirements for personal care robots and identified several gaps related to the definition of safety, identification of hazards, risks, and emerging challenges related to SOTIF. In their work they argue a need for a new robotic system specific safety standard that manages these limitations. Authors (Roberto Pietrantuono and Stefano Russo, 2018) discuss software engineering, organizational, cultural and educational issues as the main limitations for safety. Associating safety conformance

---

[1] Coverage in the context of conformance to machine safety regulatory requirements refers to the extent to which a risk/hazard analysis covers all reasonably foreseeable safety-critical situations.

[2] Traceability in the context of conformance to machine safety regulatory requirements refers to the extent to which a safety requirement can be traced from establishment to integration, to validation and verification of its availability.

limitations in the robotics domain with quality software engineering is further corroborated by (Reichardt et. al., 2013), stating that the use of probabilistic and data-driven components significantly increases the complexity of large software-based systems.

We expect that as the level of automation in machinery increases similar safety conformance limitations emerge. This paper thus discusses limitations specific to heavy machinery. After we introduce EU Machine Safety regulations in Section 2, we discuss our main contribution. We then identify emerging conformance limitations and provide a concrete example in Section 3. Conclusions are discussed in Section 4.

2 EU machine safety conformance

Any justification towards a machine's level of safety rests on three core principles: coverage[1], traceability[2] and explainability[3]. The manufacturer must demonstrate that all reasonably foreseeable hazards and risks have been covered by safety functions, such functions must be traceable from their requirements to their implementation and a justification towards their availability must be properly explained as guided by the Machinery Directive and therein referenced standards (EU commision; Machinery Directive, 2006).

Increasing the level of automation embedded in machinery limits machine safety conformance owing to the fact that it reveals shortcomings in establishing a justifiable argument based on the three core principles of certifiable machine safety. Sufficient coverage[1] is justified through the application and documentation of a risk assessment and hazard analyses as guided by relevant standards, increasing the level of automation uncovers risks and hazards currently not considered by the machine safety standards. Traceability[2], in turn, reflects the level in which a safety requirement can be traced within the machine, increasing automated functionalities adds to the complexity of interrelated components complicating documented proof of a safety functions continuous availability. Explainability[3] is established at a documentational level combining both coverage and traceability in such a manner that third parties[4] can confidently state that a machine conforms to machine safety requirements.

At present, Machine safety is designed and developed in such a way that the principles of the Machinery Directive are evaluated and safety-critical situations

---

[3] Explainability in the context of conformance to machine safety regulatory requirements; refers to the extent in which a justifiable and convincing argument can be made towards the reliable availability of safety functions

[4] Third parties in the context of conformance with machine safety regulatory requirements; refers to authorised representatives and notified bodies responsible for issuing either a declaration of conformance or a certificate of certification.

which occur from the machines intended use and potential misuse are mitigated through embedding relevant safety functions. A manufacturer performs a risk assessment (12100:2010, 2010) and hazard analysis (14121-2:2012, 2012). From which emerging safety requirements are translated into safety functions and matched to a required performance level (PL_r) (13849-1:2015, 2015) or system integrity level (SIL) (62061:2005, 2005). This determines the level of engineering rigour required for the successful implementation of a safety function. These safety functions must then be consolidated into an appropriate architecture embedding each safety function into an electrical control systems up to their PL_r or SIL. Lastly, throughout this process the manufacturer is expected to document each phase of the design, development, integration and validation of safety supporting the manufacturer in a justification towards the three core principles of certifiable safe machinery. Table 1 shows an overview of machine safety standards typically applied by manufacturers in the process of machine safety conformance. The standards referenced here are harmonised with Machinery Directive 2006/42/e, meaning they rest under an assumption of conformity with this directive.

Table 1. Harmonised machine safety standards guiding the machine safety conformance process.

Reference number	Title	Type
EN ISO 12100:2010	Safety of machinery - General principles for design - Risk assessment and risk reduction	Type A[5]
EN ISO/TR[6] 14121-2:2012	Safety of machinery — Risk assessment — Part 2: Practical guidance and examples of methods	Type A[5], Technical report
EN ISO 13849-1:2015	Safety of machinery - Safety-related parts of control systems - Part 1: General principles for design	Type B[5]
EN ISO 13849-2:2012	Safety of machinery - Safety-related parts of control systems - Part 2: Validation	Type B[5]
EN 62061:2005	Safety of machinery - Functional safety of safety-related electrical, electronic and programmable electronic control systems	Type B

---

[5] Type A standards define basic concepts, principles for design and general aspects applicable to machinery. Type B standards define one safety aspect or one type of safeguard that can be used across a wide range of machinery.

[6] A technical report (TR) contains data included from a survey or informative report or provides information of the perceived "state of the art".

2.1 Changes in machine safety standards, brought on by automation

Forthcoming replacement of the Machinery Directive 2006/42/ec by a Machinery regulation further complicates machine safety conformance for manufacturers. While drafts of this regulation have been made available, it remains unclear exactly when the Machinery Directive will be superseded by this Machinery Regulation. However, it is expected that a transitional period will be provided to manufacturers (EU commission: Machinery regulation proposal, 2022). A significant change drafted in the Machinery Regulation is the inclusion of software in the list of safety components. As a result, software that fulfils a safety function and is placed on the market, must be issued an EU declaration of conformity.

Meanwhile the international organisation for standardisation (ISO) is making headway in the development of machine specific safety standards as the level of automation continues to increase such as ISO 17757:2019 Earth-moving machinery and mining — Autonomous and semi-autonomous machine system safety, ISO 18497:2018 Agricultural machinery and tractors — Safety of highly automated agricultural machines — Principles for design and ISO 3691-4:2020 Industrial trucks — Safety requirements and verification — Part 4: Driverless industrial trucks and their systems do address machine type and application specific challenges for safety as the level of automation increases. The application areas addressed by these standards are highly machine type specific and they all still consider the human operator to be ultimately responsible for safety, limiting their applicability in machinery with higher levels of automation.

As the level of automation embedded into machinery continues to increase, we believe it is important to provide manufacturers with appropriate guidance in justifying such machine's level of safety. In section 3, we will discuss these limitations as they emerge during the machine safety design and development process for which we need appropriate standards guiding the mitigation and avoidance process of newly emerging safety hazards.

3 Emerging safety conformance limitations of highly automated and autonomous machinery

From the automotive domain, we have learned that moving towards a fully autonomous machine -one where we no longer solely rely on human input- is a stepwise process. These steps are documented in SAE J3016 an automotive standard describing the evolution of automation through the stepwise integration of automated and autonomous functionalities (J3016, 2019). Nonetheless, this standard cannot directly be applied to heavy-duty mobile machinery because hazards and critical situations which emerge during the operation of such equipment differs from those covered by SAE J3016.

In an effort to bridge this gap, the authors in (Machado et. al., 2021) propose a taxonomy of levels of automation for working machines. Inspired by the automotive standard SAE J3016 (J3016, 2019), it describes the levels of automation for heavy-duty mobile machinery in 5 levels for driving and 5 levels for manipulation in a two-dimensional matrix form. Starting at level 0 for no automation, moving upwards to level 5 for a fully autonomous machine. See Figure 2 for a visual representation of this matrix. The dashed line in Figure 2 represents a visual point where the scope of existing machine safety standards is exceeded as the level of automation transcends into level 3 "Conditional automation". In section 3.1, we proceed to clarify how increasing the level of automation exceeds the scope of guidelines provided by the standards, revealing conformance limitations with the Machinery Directive 2006/42/ec and standards mentioned in Table 1.

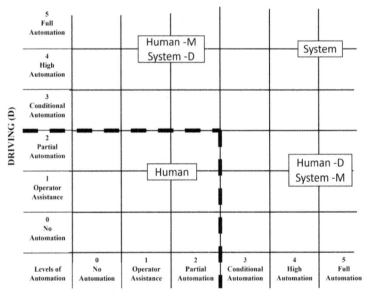

Fig. 2. The proposed taxonomy of automation by (Machado et. al., 2021). The dashed line indicates where the level of automation exceeds the scope of Machinery Directive 2006/42/ec

3.1 Safety development lifecycle discussion of emerging limitations through a concrete example

Consider, the safety development lifecycle of an excavator whose level of automation transcends level 2 "partial automation" into level 3 "Conditional automation". A simplified diagram of this machine's safety development lifecycle is shown in Figure 3, derived from sections 4.6 of EN 13849-1:2015 (13849-1:2015, 2015) and section 6 of EN IEC 62061:2005 (62061:2005, 2005).

In *phase 1 "Identification of safety functions to be performed by the SRP/CSs"*, safety functions are derived from a list of safety requirements established through a risk/hazard assessment analysing potential failures as guided by EN ISO 12100:2010 (12100:2010, 2010) and EN ISO 14121-2:2012 (14121-2:2012, 2012). Analysing an excavator (see Figure 1) shows that safety-critical situations can emerge at any level of automation e.g. erroneous bucket movements, collisions with machinery, humans and objects in the environment. At zero levels of automation the operator is in full control of the machine making him responsible for safety. The risk of erroneous operations leading to safety-critical situations is sufficiently reduced through a justifiable reliability of the machine's safety functions, proper training of operators and the availability of

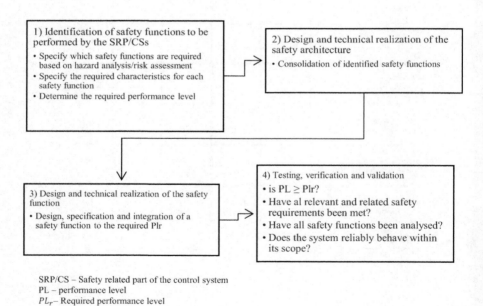

SRP/CS – Safety related part of the control system
PL – performance level
PL_r – Required performance level

Fig. 3. The four phases of a machine's safety design/development process.

manuals as stated in the Machinery Directive Annex I (EU commision; Machinery Directive, 2006). Raising the machines automated capabilities to level 2 "partial automation", the operator remains in direct control of the machine and remains responsible for the situational awareness required to ensure safety. The operator is always needed to initiate any automated functionalities and is expected to intervene when necessary. Here further risk reduction can take place through software-based safety functions automatically blocking any operations where the machine could end up in safety-critical situations. Such a safety function would solely rely on the reliability of its individual proprioceptive components, and as such does not necessarily exceed the scope of current Machine Safety legislative requirements. It does implement a logic unit that ensures a safety function and therefor requires separate assessment through a notified body as per Machinery Directive Annex IV (EU commision; Machinery Directive, 2006).

We run into compliance issues when the level of automation in the machine transcends into level 3 "conditional automation". The machine -given all task conditions are met- is capable of executing automated functionalities based on its own discernment. This ability to discern the safety of executing particular functionalities relies heavily on the machine's ability to perceive itself and the environment it operates in, increasing the risk of safety-critical situations. Thus, increasing the requirements for safety to such a level we exceed the scope of existing Machine Safety legislative requirements.

Presently, the machine safety standards define two types of machinery related faults influencing the availability of a safety function e.g. random and systematic faults. The rising level of automation leads to new failure types not considered by the existing standards, this in turn contributes to insufficient coverage[1]. Guidelines in machinery standards should expand to cover emerging hazards and risk stemming from *run-time failures* and *algorithmic failures*. One method could be via a standardised guidance on the development of safety cases along the lines of that provided in UL4600. A safety case as described in UL4600 can be viewed as the process of systematically removing different forms of uncertainty in a system. Following such a safety-case approach is a well-established method for the systematic mitigation/avoidance of risks which emerge as the level of automation increases (Vandenberg, 2019). However, we believe more research is needed on further explicitly defining these emerging failure types and relevant mitigation/avoidance methods.

o *Run-time failures* – When safety-critical situations emerge due to functional insufficiencies related to a safety function i.e. machinery specific SOTIF. For example, when an excavator encounters loose sand, or runs on a ramp which is too steep -and is incapable of appropriately adapting to these changing surface conditions- the risk of a safety critical situation such as the machine toppling over increases.

o   *Algorithmic failures* – At automation level three reliable perception plays a key role in establishing a sufficient level of safety. For this the system must integrate probabilistic and data-driven algorithms, these lead to emerging hazards currently not properly identified by the standards. Furthermore, best-practices for training and development in addition to mitigation methods for handling both potential failures and inherent weaknesses of such algorithms, remain unaddressed. For example, a false negative obstacle detection leads to a dangerous collision.

In *phase 2 "Design and technical realisation of the safety architectural design"*, safety functions must be consolidated in a verifiable architecture, handling the complexity stemming from the interrelation between multiple hardware, deterministic, probabilistic and data-driven software-based safety related components.

o   *Convoluted architectural design patterns* – Large software systems are inherently complex, and if not properly designed can lead to a loss of overview through which design flaws that could pose safety-critical situations can emerge withing the design. As the amount of software related functionalities grows so does the complexity of the system executing them. For instance a level three excavator requires a significant amount of software ensuring its automated functionalities are executed. Safety-critical situations occur when a function is not executed in a proper sequence, or the execution of a safety function is blocked by the execution of other functions. For example, instead of executing a breaking operation the machine does not interrupt its current driving operation increasing the risk of an undesired collision, or, while the bucket is scooping up materials the machine interrupts this operation via a sudden driving movement towards a different direction. Such architectural insufficiencies can lead to death or serious injuries to humans, loss or severe damage to the machine or environmental harm.

In *phase 3 " Design and technical realisation of the safety function"*, each element of a safety function is integrated into the safety-framework established in the previous phase. The Machinery Directive and standards for Machine Safety (See Table 1) currently assume a level of determinism and traceability from requirement to integration in safety functions. However, neither a significant level of deterministic behaviour nor traceability is present in probabilistic and data-driven components. A lack of normative guidelines on data-driven intended behaviour and quality integration practices of such probabilistic and data-driven components contributes to the increased risk of undesired behaviour and therefore challenges certifiability. Guidelines could focus on establishing pattern-like solutions such as wrappers, harness and workflows to embed probabilistic and data-driven elements as discussed by (Mikkonen, 2021).

o   *Data-driven intended behaviour* – Datasets used for training and validation purposes must sufficiently cover the usage domain, its intended operations, where possible be prepared for adversarial attacks and have safety measures in place for scenarios occurring outside of the defined domain. For example, a safety function is specifically trained for the detection of grey concrete obstacles in the vicinity of the excavator, the existence of a black concrete obstacle might go undetected. However, both are obstacles, and both pose a collision hazard.

- *Quality integration* – Increasing the amount of software-based safety-related components significantly increases the inherent complexity. As many different components are required to seamlessly interact with each other quality integration becomes a significant factor in increasing the reliability of a system. For example, embedding sufficient situational awareness requires the fusion of data from multiple sources, however as the machine operates in real-time its safety-related behaviour must be executed in real-time. Consider, for example, that the processing of data detecting an obstacle takes too long and a collision takes place.

In *phase 4 "Testing, verification of PL_r and validation of requirements"*, the manufacturer is required to document proof of reliability of systems safety functions. This proof is provided through well documented testing, verification and validation practices guided by the Machinery Directive and relevant standards such as those referenced in Table 1.

- *Formal verification limitations* – formal verification is proving or disproving the functional correctness of an intended algorithm with respect to its requirements using formal methods (e.g., abstract state machine) or mathematics (Bjesse, 2005). However, proofing the functional correctness of probabilistic and data-driven components has proven to be challenging as described by the authors in (Luckcuck, 2019). Therefore, the use of probabilistic and data-driven components and the subsequent requirement of formally verifying its reliability limits certifiability.

Reaching machine safety compliance of heavy-duty mobile machinery is achieved through bridging the three core principles of the machine safety standards i.e. coverage[1], traceability[2] and explainability[3]. Reflecting back on the clarification given previously, we assert that to achieve compliance -as the level of automation increases- we must overcome those six limitations occurring throughout the safety development lifecycle i.e. i) run-time failures, ii) algorithmic failures, iii) convoluted architectural design patterns, iv) data-driven intended behaviour, v) quality integration and vi) formal verification limitations.

4 Conclusion & future work

Manufacturers must comply with guidelines set in the Machinery Directive 2006/42/ec and therein referenced requirements to market their machinery within the EEA (EU Manufacturers, 2022). Transcending the level of automation of machinery into level 3 "Conditional automation" reveals compliance limitations with Machinery Directive 2006/42/ec and Machine safety standards (Table 1). This is further exasperated due to the upcoming replacement of the Machinery Directive 2006/42/ec by a Machinery Regulation where software is considered a safety component. As a result, software which fulfils a safety function and is placed on the market must be issued an EU declaration of conformity.

At present a manufacturers argument towards a machine's safety compliance rests on a justifiable argument established with guidance provided by the standards in Table 1. This argument embodies the three core principles of compliance within machine safety i.e. coverage[1], traceability[2] and explainability[3]. Transcending the level of automation from level 2 "Partial automation" into level 3 "Conditional automation" limitations with compliance emerge throughout the safety development lifecycle. Alongside a clarifying example provided in 3.1 we expose six limitations emerging throughout the safety development lifecycle i.e. i) run-time failures, ii) algorithmic failures, iii) convoluted architectural design patterns, iv) data-driven intended behaviour, v) quality integration and vi) formal verification limitations. We assert that reaching future compliance of highly automated and autonomous heavy-duty mobile machinery is achieved by overcoming the aforementioned limitations.

The first core principle of compliance -coverage[1]- relates to the extent to which a risk/hazard analysis covers all reasonably foreseeable safety-critical situations. At present, manufacturers cannot justify sufficient coverage of higher level automated heavy duty mobile machinery through adherence to the minimum safety requirements given by the Machinery Directive 2006/42/ec. We believe that for manufacturers to justify sufficient coverage[1] of safety requirements both the Machinery Directive 2006/42/ec and Machine Safety standards presented in Table 1, must expand to include risks and hazards related to *run-time failures* and *algorithmic failures*. One method could be via a standardised guidance on the development of safety cases along the lines of that provided in UL4600.

The second core principle of compliance -traceability[2]- which refers to the extent to which a safety requirement can be traced from establishment to integration, to validation and verification of its availability. The lack of traceability and deterministic behaviour in probabilistic and data-driven elements paired with insufficient normative guidelines on their intended behaviour along with quality integration practices limit manufacturers in demonstrating the continuous reliable availability of safety functions. Guidelines could focus on establishing solutions such as wrappers, harness and workflows embedding probabilistic and data-driven elements as discussed by (Mikkonen, 2021).

Lastly, the third core principle of compliance -explainability[3]- relates to the extent to which a justifiable and convincing argument can be made towards the reliable availability of safety functions. Convoluted architectural design patterns brought about by the need for numerous sensors, actuators, deterministic, probabilistic and data-driven elements can contain hidden design flaws leading to death or serious injuries to humans, loss or severe damage to the machine or environmental harm. At present neither the Machinery Directive 2006/42/ec nor those Machine Safety standards referenced in Table 1 discuss the adoption of alternative design patterns capable of providing a level of explainability to such large and complex designs. Guidelines could focus on providing manufactures with pattern-like solutions for the development of safe architectural designs appropriately managing these complexities.

Future work. Our general aim is to forge ahead on the path to safe highly automated and autonomous mobile machinery capable of conforming to EU legislative requirements. Therefore, future work will first analyse emerging safety requirements specifically related to heavy-duty mobile machinery. Then proceeding with a consolidation of these safety functions into a safe architectural design pattern.

Acknowledgements
We would like to thank Jyrki Sauramäki and Petri Aaltonen for our thorough discussions on this topic, and their comments on this work. This research is funded through the Doctoral School of Industry Innovation (DSII).

References

12100:2010, I. (2010). Safety of machinery. General. ISO.
13849-1:2015, I. (2015). Safety of machinery. Safety related parts of control systems. Part 1: General design principles. ISO. Retrieved from https://www.iso.org/standard/69883.html
14121-2:2012, E. I. (2012). Safety of machinery — Risk assessment — Part 2: Practical guidance and examples of methods. ISO.
21448:2022, I. (2022). Road vehicles — Safety of the intended functionality. ISO.
26262-1:2011, I. (2011). ISO 26262-1:2011 Road vehicles — Functional safety — Part 1: Vocabulary.
62061:2005, I. (2005). Safety of machinery - Functional safety of safety-related control systems. IEC. Retrieved from https://webstore.iec.ch/publication/59927
Aude Billard and Pericle Salvini, ,. (2021). On the Safety of Mobile Robots Serving in Public Spaces: Identifying gaps in EN ISO 13482: 2014 and calling for a new standard. ACM Transactions on Human-Robot Interaction (THRI), 1-27.
Bjesse, P. (2005). What is formal verification? ACM SIGDA Newsletter.
Brending et. al., ,. (2017). Certifiable Software Architecture for Human Robot Collaboration in Industrial Production Environments. IFAC-PapersOnLine Volume 50, Issue 1,, 1983-1990.
CE Marking for the machinery industry. (2022, August 2). Retrieved from Europa.eu: https://op.europa.eu/
EU commision; Machinery Directive. (2006). Retrieved from Machinery directive 2006/42/EC of the european parlement and of the council of may 2016 on machinery, and amending directive 95/16/EC (recast): Europe.eu
EU commission: AI regulation proposal, ,. (2021). Harmonised rules Artificial intelligence (Artificial intelligence act) and amending certain union legislative acts. European Union.

EU commission: Machinery regulation proposal, ,. (2022). REPORT on the proposal for a regulation of the European Parliament and of the Council on Machinery products. European Union.

EU Manufacturers. (2022, August 8). Retrieved from Europa.eu: https://ec.europa.eu/

EU Safety in the Automotive Sector. (2022, August 2). Retrieved from Europe.eu: https://ec.europa.eu/

J3016, S. (2019). Taxonomy and Definitions for Terms Related to Driving Automation Systems for On-Road Motor Vehicles. SAE committee.

Jackson et.al., ,. (2019). Certified Control: An Architecture for Verifiable Safety of Autonomous Vehicles.

Luckcuck, M. a. (2019). Formal specification and verification of autonomous robotic systems: A survey. ACM Computing Surveys (CSUR).

Machado et. al., ,. (2021). Towards a Standard Taxonomy for Levels of Automation in Heavy-Duty Mobile Machinery. In Fluid Power Systems Technology. American Society of Mechanical Engineers.

Machado et.al., ,. (2021). Autonomous Heavy-Duty Mobile Machinery: A Multidisciplinary Collaborative Challenge. In IEEE International Conference on Technology and Entrepreneurship (ICTE) (pp. 1-8). IEEE.

Philip Koopman and Michael Wagner, ,. (2016). Challenges in autonomous vehicle testing and validation. SAE International Journal of Transportation Safety, 15-24.

Philip Koopman and Uma Ferrell, ,. (2019). Computer Safety, Reliability, and Security : A Safety Standard Approach for Fully Autonomous Vehicles. Springer.

Reichardt et. al., ,. (2013). On Software Quality-motivated Design of a Real-time Framework for Complex Robot Control Systems. Electronic Communications of the EASST.

Roberto Pietrantuono and Stefano Russo, ,. (2018). Robotics Software Engineering and Certification: Issues and Challenges. IEEE International Symposium on Software Reliability Engineering Workshops (ISSREW). Memphis, TN, USA: IEEE.

SAS. (2022, August 2). European training network for safer autonomous systems. Retrieved from etn-sas.eu: https://etn-sas.eu/

Serban et. al., ,. (2018). A Standard Driven Software Architecture for Fully Autonomous Vehicles. IEEE.

Safe, Ethical and Sustainable: Framing the Argument

John A McDermid[1], Simon Burton[2], Zoe Porter[1]

[1]Assuring Autonomy International Programme

University of York, UK

[2]Fraunhofer IKS

Munich, Germany

Abstract *The authors have previously articulated the need to think beyond safety to encompass ethical and environmental (sustainability) concerns, and to address these concerns through the medium of argumentation. However, the scope of concerns is very large and there are other challenges such as the need to make trade-offs between incommensurable concerns. The paper outlines an approach to these challenges through suitably framing the argument and illustrates the approach by considering alternative concept designs for an autonomous mobility service.*

1 Introduction and motivation

In many domains, e.g. aviation, there have been long-term improvements in safety. This can be seen as a desirable consequence of consumerism (McDermid et al 2022) amongst other forces. There is now a growing interest in other characteristics of systems such as their environmental and ethical impacts. This is, in part, because safety is often now seen as a given and because the greater scope and capability of modern systems introduces other concerns, such as unfair distribution of risk amongst stakeholders. To reflect this wider set of concerns, we have previously suggested that there is a need to design and assure systems, so that they are 'safe, ethical and sustainable' (McDermid 2022). But this is easier said than done.

© JA McDermid, S Burton, Z Porter 2023.
Published by the Safety-Critical Systems Club. All Rights Reserved.

The aim of this paper is to indicate how we might achieve this goal and to demonstrate that we have done so. There are several challenges here. First, the set of concerns encompassed by 'safe, ethical and sustainable' is vast – water quality, consumption of rare-earth elements/metals, deforestation, ozone depletion to mention but a few of the possible environmental (sustainability) issues. Second, the systems can be very complex, including use of artificial intelligence (AI) so analysing them and their impacts (consequences) is a major undertaking. Third, there may be conflicts between, say, safety and ethical treatment of specific stakeholder groups. There is a trade-off here between utilitarian safety aimed at maximising safety over an entire group (e.g. all road users) against ensuring that each individual has an equal right to safety. This thinking can be extended further, e.g. technologies that improve road traffic safety, but rely on rare earth metals may lead to damage to the environment and exploitation of emerging economies thus harming future generations. Fourth, the set of concerns are often incommensurable in the sense that the units for measuring safety, e.g. risk or rate of occurrence of accidents, are very different from those for, say, air quality which has several measures, e.g. particulate density; similarly, measures for mental and physical harms are also quite different.

The approach to those challenges proposed here is to consider systems at the level, or stage, of their conceptual design and to employ argumentation to illuminate and justify the trade-offs between different system properties – but doing so depends on 'framing the argument' to reduce the potential concerns to a feasible set.

The rest of this paper is structured as follows. Section 2 considers the potential sets of concerns encompassed within 'safe, ethical and sustainable' and takes a first step towards bounding the scope of analysis. Next, section 3 draws on work on governance of complex systems (Burton et al 2021) as a way of 'framing the argument' which is later articulated by drawing on work on ethical assurance arguments (Porter et al 2022). Section 4 uses provision of mobility as a service (MaaS) as an illustration, including addressing different system attributes and stakeholder concerns within the argumentation framework. Conclusions are presented in section 5.

2 Life, the universe and everything

Douglas Adams famously said that '42' was the answer to 'life, the universe and everything' (Adams 1982). In saying that systems should be 'safe, ethical and sustainable' we are perhaps considering a slightly narrower problem – albeit one which doesn't seem to have such a simple answer! As a step towards framing the argument for 'safe, ethical and sustainable' we first consider some very broad articulations of concerns. More specifically, for this audience we take the meaning of safety to be understood and expand more on ethical principles and sustainability before considering a synthesis of the two sets of concerns. Finally, we discuss one area where we need to 'dig deep' – rare earths. As well as being a concern in itself, this illustrates the interdependencies which need to be understood when making trade-offs.

2.1 Ethical principles and artificial intelligence

In practical ethics, it is common to distinguish consequentialist approaches which focus on outcomes, and which take the right actions to be those that bring about the best consequences, and deontological approaches, which focus on duties, and which take right actions to be those that conform to some rules, e.g. do not steal. It is quite hard to relate to such concerns in the context of system design and development. However, as there have been some egregious cases of unethical consequences with systems based on AI and machine learning (ML), in recent years numerous public and private sector organisations have released sets of ethical principles for the development and deployment of AI (Fjeld et al 2020; Jobin, Ienca & Vayena 2019).

In a previous analysis (Porter et al 2022) we have drawn on insights in the ethical AI literature (Floridi and Cowls 2021) to argue that the content of these many sets of ethical principles for AI – which cover values such as fairness, safety, well-being, and privacy, to name but a few - lends itself to a more simplified framework of four ethical principles which have their origin in biomedical ethics (Beauchamp and Childress 1979). These four principles are: beneficence; non-maleficence; respect for personal autonomy; and justice. The four principles can be adapted to the ethical concerns about advanced technologies, when supported by transparency of the assurance argument as well as the ML elements themselves. Our understanding and adaptation of the four principles in the AI context is as follows:

1. beneficence (the system should bring benefit to stakeholders);
2. non-maleficence (the system should avoid unjustifiable harm to stakeholders either directly or indirectly, e.g. via environmental effects);
3. respect for human autonomy (stakeholders should have appropriate and meaningful control over the system or its impact on them);
4. justice (there should be a fair or equitable distribution of benefit and risk from the system across stakeholders).

This underpins the framing of the argument in section 3.

2.2 Sustainable development

The UN General Assembly adopted the 17 Sustainable Development Goals (SDGs) during their 70th session in 2015, as illustrated in Figure 1.

Fig. 1. Sustainable Development Goals

Some of these are essentially individual concerns, e.g. 2 & 3; some are societal, e.g. 8 & 9; many are about the environment and focus more directly on sustainability of the planetary ecosystem. The UN itself prompts international action on these goals, e.g. in October 2022 they focused on goal 2, zero hunger (UN 2022), and other organisations, e.g. UNESCO, link some of their activities to the SDGs.

But if we are designing a particular system, how can we address such global concerns? More narrowly, is it possible to work out which concerns our system might impact? The second question is easier to answer. For example, for an urban autonomous transport system, perhaps the primary SDG is 11 (sustainable cities and communities), and others such as 10 (reduced inequalities) would apply in terms of access to transportation. Some of the SDGs, e.g. 12 (responsible consumption and production) apply to almost all systems. We consider the first question in section 3 under framing.

2.3 The doughnut economy

Perhaps surprisingly, some of the thinking that has tried to blend the perspectives of sustainable development and the (ethical) needs of individuals has come from economists. The ideas are credited to Kate Raworth (Raworth 2017) and are set out in Figure 3, although the simplified 'doughnut' (Wiedmann et al 2020) in Figure 2 is a better starting place for understanding the concepts.

Fig. 2. The Doughnut Economy: Simplified View (Wiedmann et al 2020)

The basic idea is that the acceptable – sustainable, safe and just – space for humanity (including our dependence on the planet) is in the green ring[1]. Outside this ring there are environmentally unsustainable consequences, e.g. through global warming. In the centre there is unacceptable poverty and other unmet needs, e.g. lack of clean water. Taking a radial slice through the figure, this can be seen as analogous to the ALARP (As Low As Reasonably Practicable) concept – but with two ALARP triangles, one where the outside is unacceptable environmentally and the inside is unacceptable from an individual or societal perspective.

Figure 3 shows a more detailed version of the model, with subdivision of the inner circle and outer annulus into more specific concerns. Elements of the inner circle can be seen to correlate with the SDGs e.g. gender equality (SDG 5) and water (SDG 6). There are similar, although less obvious, correlations between the outer annulus and the SDGs, e.g. air pollution relates to climate action (SDG 13). But one of the reasons for showing this more detailed figure is to illustrate the difficulty of identifying and agreeing even the top-level concerns.

---

[1] This is referred to as 'doughnut economics' although the acceptable, green part, of the Figure is an annulus not a torus.

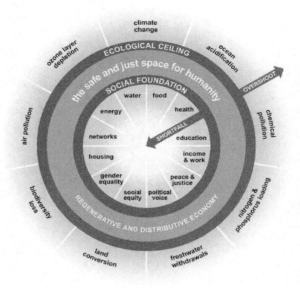

Fig. 3. The Doughnut Economy (Raworth 2017)

2.4 Rare earths

The above discussion is all high-level – it is time to come down to (rare) earth(s). Affordable and clean energy (SDG 7) depends on renewable energy sources, e.g. solar and wind, both of which are inherently variable. Also, sustainable cities (SDG 11) need sustainable energy and batteries for electric vehicles (EVs) – unless we design '15-minute cities'[2] – and this is perhaps only implicit in the SDGs (e.g. 9, 10 and 11). Such dependence on renewables is even less obvious in the doughnut economy (energy, income and work, and social equity perhaps). So, as well as the high-level structures given by the SDGs and doughnut economics we need to 'dig down' into the supply chain, including considering the rare earths needed for renewables, batteries, etc.

An analysis of what would be required to replace fossil fuels for power generation and for transport (Michaux 2021) identified significant problems with materials including rare earths, see Figure 4[3]. In several cases, the estimate is that it will take millennia at current extraction rates to produce enough materials to replace fossil fuels. Note that this is for just one generation of systems, e.g. EVs or wind farms, and does not consider replacements – where reuse/re-cycling would be vital.

---

[2] Where all citizens' needs can be met within a 15-minute walk or cycle ride from their homes.

[3] The figure comes from an associated presentation (Michaux 2022), not the report itself.

Metal	Element	Total metal required produce one generation of technology units to phase out fossil fuels (tonnes)	Global Metal Production 2019 (tonnes)	Years to produce metal at 2019 rates of production (years)
Copper	Cu	4 575 523 674	24 200 000	189,1
Nickel	Ni	940 578 114	2 350 142	400,2
Lithium	Li	944 150 293	95 170 *	9920,7
Cobalt	Co	218 396 990	126 019	1733,0
Graphite (natural flake)	C	8 973 640 257	1 156 300 ♦	3287,9
Graphite (synthetic)	C		1 573 000 ♦	-
Silicon (Metallurgical)	Si	49 571 460	8 410 000	5,9
Vanadium	V	681 865 986	96 021 *	7101,2
Rare Earth Metals				
Neodymium	Nd	965 183	23 900	40,4
Germanium	Ge	4 163 162	143	29113,0
Lanthanum	La	5 970 738	35 800	166,8
Praseodymium	Pr	235 387	7 500	31,4
Dysprosium	Dy	196 207	1 000	196,2
Terbium	Tb	16 771	280	59,9

Fig. 4. Materials Needs for Replacing Fossil Fuels (Michaux 2022)

Further, there are environmental impacts of mining – for example it is suggested that 15 tonnes of CO_2 are emitted for every tonne of lithium mined (Crawford 2022). The key points here are that there may be fundamental difficulties in meeting some of the SDGs and that the range of concerns for the 'doughnut economy' implicitly involves some conflicts. Thus, a complete argument needs to encompass the effects of the supply chain, including mining metals and rare earths, not just the fundamentals of human life such as air and water. Put another way, as well as considering overall goals we need to be mindful of the constraints in meeting those goals and their interdependencies which we will need to understand and respect when making trade-offs.

3 Framing the argument

The discussion above is mainly on a broad scale, e.g. concerns affecting nations and, in some cases, relates to impacts on a planetary scale. So, what do engineers do for a particular system of concern? Systems engineers normally consider trade-offs across a range of factors – although not as broad as we are proposing here. This raises questions of skills and competence. There are also questions about the extent to which decisions will be taken at a political level. These are complex issues which we can only partially address, but we return to them in the conclusions.

Here, we focus onto those concerns that: a) the system design can impact, and b) are in the scope of control or influence when designing the system. This enables us to establish an appropriate framing of the concerns. We approach this by first considering how to govern the safety of complex systems and second how to articulate arguments about the (ethical) acceptability of an individual system.

3.1 Safety of complex systems

The SDGs and many of the concerns in the 'doughnut economy' are at very broad scale and not something that can be controlled by any one system or development. However, if these concerns are not considered in developing or deploying individual systems then it is very unlikely that these goals will be met. A study of complex systems which exhibit emergent properties, e.g. effects of CO_2 emissions on the environment, proposed identifying safety controls at three layers – task & technical, management and governance, see Figure 5 (Burton et al 2021). We can broaden this model to cover safe, ethical and sustainable behaviour – and identify what concerns can and should be addressed at each layer.

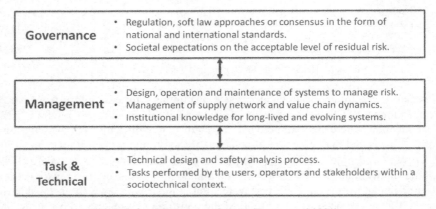

Fig. 5. Safety Management Layers (Burton et al 2021)

The critical decision-making that shapes systems and their wider impacts take place at the *management* layer – choice of system concept (and then the architecture, materials, capability, etc.) can all have a very broad effect. But an individual company making, say, an EV has no control over how electricity is generated (for their production or for individual vehicle owners), they do not dictate the policies for where rare earths are mined, etc. Thus, *governance* (let's say at the level of a nation) must put in place laws, policies, incentives, etc. that shape managerial decision-making. For example, this might be in terms of a maximum life-time carbon footprint for an EV together with policies for recycling critical materials. In this context, work at the *task & technical* layer mainly provides information on which management level decisions can be made, see the discussion on ethical assurance arguments below and the illustrative example in section 4.

3.2 Ethical assurance arguments

The use of safety cases is well-established, if not always well-practised. We have previously proposed extending the notion of safety cases to ethics – thus producing ethical assurance arguments (Porter et al 2022).

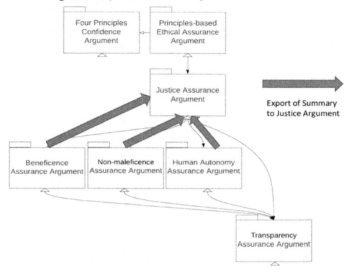

Fig. 6. Modular Structure of the Ethical Assurance Argument (Porter et al 2022)

The ethical assurance argument (see (Porter 2022) for more details) builds on concepts from biomedical ethics, as introduced above, in particular addressing:

- Beneficence – providing benefit from the system
- Non-maleficence – freedom from harm from the system (this includes safety considerations, i.e. avoidance of harm to individuals)
- Human autonomy – the ability to have appropriate human control over the system, including the ability to make meaningful choices about being affected by it, e.g. to opt out of its use
- Justice – the balance of benefit and risk from the system, including any constraints on human autonomy, across stakeholders, e.g. that benefits do not accrue to one set of stakeholders with the (risk of) harms falling on a different (disjoint) set of stakeholders.

As the concern in developing this argument structure was with AI/ML-based systems, and as ML models are often highly complex 'black boxes', demonstrating that the goals above have been met will often require the use of techniques that provide algorithmic visibility. Transparency is a supporting principle in the ethical assurance argument covering the ML models, and transparency of the assurance argument itself (i.e. visibility of the reasons and evidence for specific claims).

In using the argument framework, summaries of the argument in each of the four lower-level argument modules are exported to the justice argument. The approach can be adapted to deal with the wider safety, ethical and environmental concerns – safety is already addressed under non-maleficence and environmental concerns can be considered as both benefits and harms, e.g. removing plastic from the oceans is a benefit, and pollution of the atmosphere is clearly a harm. The scope of human autonomy might also be expanded – for example, in many cases nations make choices on behalf of citizens, e.g. on the mix of fuels for electricity generation, with individuals limited to narrower choices, e.g. choice of energy supplier.

4 An illustration – mobility as a service

To make the above ideas more concrete we consider the introduction of mobility as a service (MaaS) into a city, focusing on the decision-making at the level of the city (management layer). We assume that the city's current public transport system uses diesel-powered buses, and the Council wants to move to providing services autonomously, using EVs to improve air quality. The illustration shows how different options (vehicle mixes) can be compared using tables to summarise the beneficence, non-maleficence and human autonomy argument modules, so that the best approach can be chosen in the justice module. It is necessarily hypothetical but uses data, e.g. on vehicle carbon footprint, which is as realistic as possible. The illustration focuses on the trade-offs but see sections 4.8 and 4.9 for a discussion of the argument.

4.1 The status quo and the ambition

The city currently has a fleet of 50 buses (single and double-decker) providing about 250,000 passenger kilometres (pa-km) of journeys a day. The buses operate for 16 hours a day and travel is free for citizens over 60 years of age, after 9am. The city centre has several pedestrianised areas, and the Council has already limited access for private vehicles except to blue-badge holders (see below for an explanation) and residents. Unusually, the city doesn't allow taxis or private-hire vehicles to operate in the city centre[4].

The Council wishes to improve air quality and to provide transport through an MaaS scheme using EVs and asks their transport planners to come up with a range of options so they can choose the favoured approach. In doing the analysis to present ideas to the Council the transport planners decide that they need to consider different stakeholders, focusing on the users of the transport system:

---

[4] Not very likely but this can be viewed as an assumption to make the illustration more compact!

- Lone traveller – an individual who would prefer to travel alone (i.e. in a sole occupancy vehicle) for their safety and peace of mind
- Large family – parents and children travelling together who would prefer not to be separated[5]
- Young family – parents (or single parent) with one or two children using a pram or pushchair
- Blue badge – individuals with mobility needs and who are allowed to travel and park in areas otherwise not accessible to traffic
- Free travel – individuals living in the city and over the age of 60 entitled to a pass giving them free travel on the city's buses (after 9am)

There are also other stakeholders, including bus drivers and vehicle maintenance staff, whose employment can be affected by the MaaS operation.

4.2 Fleet options

The Council officials decide to explore three different vehicle mixes which they can then compare – including comparing against the current services.

Table 1. Fleet Options: Numbers of each type of vehicle

Vehicle Type	Fleet A	Fleet B	Fleet C
2 seat LPV	250	125	0
4 seat LPV	250	125	0
Bus	0	25	0
Shuttle	0	0	175

Council staff have seen several electric light passenger vehicles (LPVs), for example see Figure 7, and decide that one possibility (Fleet A) would be to use a mix of 2 and 4 seat LPVs. This would enable provision of services to lone individuals without having poor utilisation in larger vehicles, and the LPVs would be available on demand and able to go to a destination pre-selected by the passenger(s). The 4-seat LPVs are flexible in use and can be configured easily, for a particular journey, e.g. to hold a 2-seat buggy in the back (by folding down the rear seats) whilst still accommodating two adults in the front and thus are suitable for young families.

---

[5] In the UK around 28% of families have 3 or more children.

Fig. 7. Sven LPV[6]

A hybrid (Fleet B) would include buses on fixed routes, as now, along with a mix of 2 and 4 seat LPVs. The buses would only operate peak hours (8am to 6pm) and have a capacity of 60. For cost reasons, only half the number of LPVs is possible by comparison with Fleet A, but analysis of the current service usage indicates that this would cater for the city's needs as the LPVs would be available 24/7.

A third option is to use shuttles – like a mini-bus – with a maximum occupancy of 8 passengers (Fleet C) which can meet the overall transportation need but only by having journeys shared between different passenger groups.

4.3 Safety (non-maleficence)

The information that needs to be made available for scrutiny to support the justice argument was outlined in section 3.2 and needs to be at whole-fleet level. The data arising from traditional safety assessments can be summarised in two tables, where the percentages in Table 2 reflect the scores against specific vehicle tests as part of the Euro New Car Assessment Programme (NCAP)[7]:

Table 2. Euro NCAP Summaries by Vehicle Type

Vehicle Type	Adult Occupant	Child Occupant	Vulnerable Road Users	Safety Assist
2 Seat LPV	35%	42%	56%	24%
4 Seat LPV	65%	72%	60%	35%
Shuttle	91%	81%	73%	81%
Bus	23%	38%	18%	15%

---

[6] See: https://www.fev.com/en/media-center/press/press-releases/news-article/article/sven-shared-vehicle-electric-native-optimized-for-urban-mobility.html (accessed 16th October 2022)

[7] See: https://www.euroncap.com/en (accessed 2nd November 2022)

Table 3. Safety Ranking by Fleet

Safety	Fleet A	Fleet B	Fleet C
Rank	2	3	1

Buses perform poorly because of their mass (and shape) in impact with vulnerable road users (VRUs). The shuttles are best, as they are full-size vehicles and can be engineered with crumple zones, safety assistance systems such as automated emergency braking, etc. The LPVs are relatively poor as they don't have the size or power to include safety assistance features, with 2-seat LPVs worse due to limited space for crumple zones. The VRU score is quite high due to the low vehicle mass.

4.4 Environmental impact (non-maleficence)

The environmental impact is estimated[8] based on the full life of the vehicles including manufacturing as well as the operation of the vehicles; this implies some assumptions about how the electricity is generated as well as the useful vehicle life.

Table 4. Life-cycle carbon footprint of each vehicle type

Impact	2 Seat LPV	4 Seat LPV	Shuttle	Bus
Vehicle	60 g/CO_2/km	90 g/CO_2/km	230 g/CO_2/km	3,000 g/CO_2/km
Occupancy	1.5	2	5	20
Per Passenger	45 g/CO_2/pa-km	45 g/CO_2/pa-km	46 g/CO_2/pa-km	150 g/CO_2/pa-km

Table 5. Life-cycle carbon footprint for each fleet, for each day of use

Impact	Fleet A	Fleet B	Fleet C
LPV pa-km	250,000	150,000	0
Bus pa-km	0	100,000	0
Shuttle pa-km	0	0	250,000
Total (kg)	11,250	21,750	11,500

Given the data in table 4 including estimates of average occupancy, the carbon footprint of the different fleet options can be assessed, see table 5, setting out carbon cost per day. Fleet C is little different from Fleet A – although the shuttles are worse than the LPVs per vehicle kilometre this is more-or-less exactly balanced out by the higher average occupancy. The buses have a much worse impact due to their size and the expected occupancy.

---

[8] The figures are loosely based on an analysis of LPVs by Zemo (Zemo 2021).

4.5 Availability of transport (beneficence)

The primary intended benefit of the MaaS is to provide (better) transport services to citizens and visitors to the city. The benefits are assessed qualitatively with N meaning there is no substantial difference from the current bus service with +/++ and -/-- denoting smaller/larger improvements and detriments, respectively.

Table 6. Availability of transport per stakeholder for each fleet

Stakeholder	Fleet A	Fleet B	Fleet C
Lone traveller	++	+	N
Large family	--	-	+
Young family	-	-	+
Blue badge	--	--	N
Free travel	++	++	+

The table reflects value judgements – for example that Fleet A has problems for young families as they won't be able to get buggies into half of the LPVs. Fleet B is judged to be similar for young families as buses (which can easily take buggies) are on fixed routes and there are fewer 4 seat LPVs available. And so on.

4.6 Availability of employment (beneficence)

The employment benefits relate both to the bus drivers and to maintenance staff. Arguably the loss of employment could be seen as a harm, but we treat employment as a benefit for the purpose of this illustration.

Table 7. Availability of employment per stakeholder for each fleet

Stakeholder	Fleet A	Fleet B	Fleet C
Bus drivers	--	-	--
Maintenance	++	+	+

There may be compensation for the bus drivers, but that is left to the discussion of the argument.

4.7 Human autonomy

Human autonomy relates to the meaningful control related to the system, including appropriate freedom of choice for each stakeholder, for example the ability of a lone traveller to ride in a single occupancy vehicle. In making the value judgements here we are assuming a level of control, for example being able to ask for a 2-seat LPV when calling for a vehicle. In the case of large and young families, the score reflects the need to wait for larger vehicles and the fact that very large families (5 or more individuals) would have to split up on a journey (with Fleet B, this would only be necessary outside core hours or if travelling far from the bus routes). Although blue badge holders, e.g. wheelchair users, can use buses and the shuttles, they would have less freedom – not being able to travel when they want and where they want, hence all the options represent a detriment.

Table 8. Autonomy per stakeholder for each fleet

Stakeholder	Fleet A	Fleet B	Fleet C
Lone traveller	++	+	-
Large family	--	-	+
Young family	--	-	N
Blue badge	--	-	-
Free travel	N	N	N

4.8 Making the argument

Ethical assurance arguments would have to be made for each concern – and sections 4.3 to 4.7 have illustrated the core data related to beneficence, non-maleficence, and human autonomy via the tables "exported" to the justice argument. In practice the arguments would be more complex, and the above should be viewed as fragmentary illustrations. Further, the arguments and evidence are subject to uncertainty and there will be a need to monitor systems in operation to collect leading indicators of risk, etc. providing continual assurance. We return to this point in the conclusions.

It is helpful to briefly consider the main argument modules before considering the justice argument. No consideration is given to transparency; it may be that these autonomous EVs use ML so transparency might be needed, for example, to estimate the VRU rating for the different vehicles. However, we view discussion of transparency, e.g. via explainability (McDermid et al 2021), as outside the scope of consideration here. There is also a wider issue of transparency of the assurance argument and evidence which we return to in the conclusions.

At the level of the three main argument modules – beneficence, non-maleficence, and human autonomy – there are two key criteria to meet. First, there are criteria which relate to the broad set of concerns identified above, e.g. the SDGs or elements from the (models of) the doughnut economy which "flow down" from the governance layer. This might, for example, be a requirement that some system is carbon neutral over its life – or, more likely in the case of the EVs in our illustration, there is some maximum carbon footprint per passenger kilometre based on knowledge that there will be a carbon offset elsewhere. If, for example, the limit was 50g/CO_2/pa-km then Fleet B would be rejected. However, at a figure of 100g/CO_2/pa-km for the fleet then Fleet B would still be open for consideration even though the buses don't meet this target.

Second, there are internal criteria within each of the modules, for example, in the human autonomy argument, one such criterion would be that individual stakeholders can give informed consent to the use of the system in a way that affects them – for example, can a lone traveller choose who gets into a shared shuttle with them and/or the route the shuttle takes in completing its journey? A judgement might be made that, without such controls, Fleet C is unacceptable due to the constraints on the person's capacity to manage their own personal safety – especially late at night.

In both these cases, this can be viewed as being like an ALARP criterion – if a threshold is met then the option can be considered, in a similar way to being in the broadly tolerable region of ALARP. Those options that "survive" the module- level can be considered in the justice argument. In the illustration here, we assume that all three options "survive", so tables 3, 5, 6, 7 and 8 are the summaries that are "exported" to the justice argument as indicated in Figure 6.

Arguably, the ideal situation is that one of the options Pareto-dominates all the others, i.e. is better with regards to all the concerns. This isn't true in this case. Fleet C is, on most measures, as good as or better than the others (the difference in carbon footprint in table 5 is small enough to ignore) but there are significant disadvantages for bus drivers (table 7) and issues for both blue badge holders and lone travellers (table 8). There are potential resolutions (mitigations) for these issues across the safety management layers introduced above, e.g.

- Bus drivers – management layer change, offering retraining for other roles
- Blue badge holders – governance layer change, altering the policy to allow blue badge holders to enter the city centre as before (perhaps with incentives to adopt EVs)
- Lone travellers – management/task & technical layer change through adopting a different fleet mix (shuttles plus some LPVs) giving options for single occupancy travel (there are also single occupancy vehicles[9])

The approach to ethical arguments (Porter et al 2022) incorporates the notion of using (wider) 'reflective equilibrium' to reach a balance between conflicting demands – put another way, justifying the trade-offs. Reflective equilibrium is most closely associated with the work of the political philosopher John Rawls (Rawls 1951; Rawls 1971). In this context, we take wide reflective equilibrium to mean the end-point of a decision process which involves stakeholders (or their trusted representatives) and other decision-makers working back and forth between their considered ethical judgements about specific competing demands, general ethical principles that apply, and relevant non-ethical judgements (e.g. technical or financial) until they reach a coherent opinion. Reflective equilibrium is achieved when none of the parties involved are inclined to revise any of their component judgements or

---

[9] e.g. the Electra Meccanica Solo https://www.emvauto.com/solo (accessed 16th October 2022).

beliefs about the decision or trade-off further because together these have the highest degree of acceptability (Daniels 2020).[10] In this illustration, that approach might be used at two stages – initially identifying issues that can't be resolved, and later assessing (and accepting) the revised fleet choice and governance changes. One way of representing this would be to record the positions of each stakeholder.

4.9 Observations

One of the challenges for the 'safe, ethical and sustainable' mantra is the wide range of concerns (e.g. from the SDGs and the doughnut economy) that might need to be considered. Based on the illustration outlined above it seems reasonable to view the principles underlying the ethical assurance arguments as a good way of structuring concerns, whereas the SDGs and doughnut economy give good prompts, but neither can be viewed as exhaustive. Consequently, there is an important role for 'reflective equilibrium' in framing the argument for each system being considered although it remains to be seen how best to represent that in the justice argument.

However, in the discussion above there has been little explicit reference to framing the SDGs or the 'doughnut economy'. In effect, this framing "falls out" from the nature of the system (or at least it does, up to a point). We can consider each of the main elements of the argument in turn and how they relate to the broader goals:

- Non-maleficence (safety) – a specific aspect of SDG 3 (good health and well-being) but not something that is obviously reflected in the doughnut economy
- Non-maleficence (environment) – SDGs 11-13 (sustainable cities and communities, responsible consumption and innovation, and climate action) plus climate change and air pollution from the doughnut economy
- Benefits (employment) – SDG 8 (decent work and economic growth) and income and work from the doughnut economy
- Benefits (availability of transport) – not directly reflected although it arguably underpins employment benefits
- Human autonomy – perhaps this is implicit in gender equality (SDG 5) and reduced inequalities (SDG 10) and similarly maybe an aspect of social equality from the doughnut economy, but it is not really explicit

In addition, the layering of controls into governance, management and task & technical, enables assignment of responsibility for managing the concerns amongst different stakeholders who should have the appropriate knowledge and authority to discharge those responsibilities. This has been at least partially illustrated above.

---

[10] For good discussions of the method of reflective equilibrium in engineering ethics, see van de Poel & Swart (2010) and van den Hoven (1997).

5 Conclusions

The idea of considering whether systems are 'safe, ethical and sustainable' might be compelling in principle, but it is not so obvious how to meet this goal. The aim in this paper has been to shed some light on how this goal might be met by drawing together ideas from sustainable development, 21st century economics, management of safety in complex systems and ethical assurance arguments.

Our view remains that the key ethical concepts of beneficence, non-maleficence, human autonomy, and justice are a good way of structuring arguments about the acceptability of systems. Whilst we have not illustrated the dynamics of a reflective equilibrium process, it seems one of the few practical "tools" to address the trade-offs necessary between incommensurable concerns. Further, the layering of controls into task & technical, management and governance seems to help identify the best locus for addressing some of the overarching concerns.

However, the scope of potential concerns is vast and even the simple illustration here shows that the widely accepted SDGs and other models such as the doughnut economy do not embrace all the concerns – in particular, neither seem to reflect the notion of human autonomy very clearly or directly. Thus, we would advocate using the argumentation approach outlined here, treating the SDGs, doughnut economics – and potentially other frameworks – as checklists to make sure concerns which are important for a given system have not been overlooked. But our expectation is that, in many cases, the framing will "fall out" from appropriate consideration of the conceptual design for a system as it did in the illustration here. Time will tell if this is a realistic expectation in more complex settings.

Nonetheless, the analysis laid out above will, in reality, be much more complex, in particular as it will be based on evidence and subjective judgement that may include a high level of uncertainty and whose validity may erode over time as the environment in which the system operates, including behavioural patterns of the users, evolves. Thus, there will be a need for "continuous assurance" including an identification of the observation points required to collect leading risk indicators such that the system can be adapted to meet evolving needs and safety, ethical and sustainability trade-offs – which might also vary across countries and cultures.

But who has the skills and authority to conduct this work? Systems engineers are already used to designing systems to meet multiple, often conflicting criteria (e.g. performance, vs. cost vs. safety vs. usability). However, it is too much to expect that a technically trained engineer would have both the competence as well as the responsibility to handle this multitude of additional concerns. This is a place where reflective equilibrium has a role to play. Such a process can involve specialists who can represent these different specialisms. There is a need for systems (or safety) engineers who are capable of co-ordinating across the relevant disciplines. It is not uncommon to talk about a 'T-shaped' engineer who has broad skills and depth in one specific area, giving them authority (from the depth) and the skills to manage a multi-disciplinary team (from the breadth). The 'safe, ethical and sustainable' mantra suggests that the breadth for some engineers needs to be into ethical and environmental concerns, not just engineering issues. This implies a need for a refined education for at least a small cadre of engineers who have leadership roles in complex projects – and perhaps they need to be 'Π-shaped' with deep skills in two areas, perhaps environmental or ethical issues to complement a technical skill.

Further, this is where the layered model of (responsibility for) safety controls comes in. The issues should be considered from a system engineering and assurance perspective but at the management and governance layers where increasing responsibility is taken, and where this responsibility includes ensuring that the layer below produces systems that support the manifold safety, ethical and sustainability goals. This implies, for example, that the governance layer should produce an ethical assurance case, e.g. for the regulation of transport systems[11]. In other words, in order to deploy advanced technologies in a truly safe, ethical and sustainable manner, their deployment and regulation needs to be consciously "engineered" against a set of clear principles and methods for achieving transparency, including of the assurance arguments[12].

Finally, this leads to the question whether or not this approach would be broadly accepted by society used to elected politicians making "popular" decisions and in a time where rational, expert judgement is often not accepted and indeed rejected as "elitism". A sensitivity to these issues is therefore required, which is where both an ethical framing and understanding of societally perceived risk is essential. It is also one of the main motivations for the idea of a "mantra" (McDermid 2022) which, if repeated often enough, might begin to shape public and political perception and behaviours.

Acknowledgments This work is supported in part by the Assuring Autonomy International Programme, funded by the Lloyds Register Foundation, and the UKRI Trusted Autonomous Systems (TAS) programme through the Assuring Responsibility for TAS (AR-TAS) project.

References

Adams D (1982) Life, the universe and everything, Pan Macmillan.
Beauchamp T, Childress J (1979) Principles of biomedical ethics, Oxford University Press
Burton S, McDermid JA, Garnett P, Weaver R. (2021) Safer Complex Systems: An Initial Framework, https://raeng.org.uk/media/4wxiazh3/engineering-x-safer-complex-systems-an-initial-framework-report-v22.pdf (accessed 16th October 2022)

---

[11] This is not entirely unprecedented. For example, a safety case was produced for the change in vertical separation minima in European air space. What we are suggesting is a stretch beyond this, but the proposal is not without some antecedents.

[12] As an example of how this might be done in an "engineered" regulatory framework, the Centre for Data Ethics and Innovation has suggested that the Safe and Ethical Operating Concept for an autonomous road vehicle should be made publicly available, see: https://www.gov.uk/government/publications/responsible-innovation-in-self-driving-vehicles/responsible-innovation-in-self-driving-vehicles#annex-a-safe-and-ethical-operational-concept-and-safety-management-systems.

Crawford I, (2022) How much CO_2 is emitted by manufacturing batteries? https://climate.mit.edu/ask-mit/how-much-co2-emitted-manufacturing-batteries (accessed 16th October 2022)

Daniels, N (2020) "Reflective Equilibrium", The Stanford Encyclopedia of Philosophy (Summer 2020 Edition), Edward N. Zalta (ed.), https://plato.stanford.edu/archives/sum2020/entries/reflective-equilibrium/ (accessed 23rd October 2022)

Fjeld J. Achten N, Hilligoss H. Nagy, A, Srikumar, M. (2020). Principled artificial intelligence: Mapping consensus in ethical and rights-based approaches to principles for AI. Berkman Klein Center Research Publication, (2020-1).

Floridi L. and Cowls J. 2021. A unified framework of five principles for AI in society. In: Floridi L. (ed.) Ethics, Governance, and Policies in Artificial Intelligence. Philosophical Studies Series, 144: 81-90. Springer, Cham

Jobin, A, Ienca, M, Vayena, E. (2019). The global landscape of AI ethics guidelines. Nature Machine Intelligence, 1(9): 389-399

McDermid JA, Jia Y, Porter Z, Habli I (2021) Artificial intelligence explainability: the technical and ethical dimensions. Philosophical Transactions of the Royal Society A. 379(2207): 20200363.

McDermid JA, Porter Z, Jia Y (2022) Consumerism, Contradictions, Counterfactuals: Shaping the Evolution of Safety Engineering, in Safer Systems: The Next 30 Years, Proceedings of the 30th Safety-Critical Systems Symposium, Safety Critical Systems Club

McDermid JA. Safe, Ethical & Sustainable: A Mantra for All Seasons? (2022) Safety Systems. 2022 Feb 10:5-10.

Michaux SP (2021) Assessment of the Extra Capacity Required of Alternative Energy Electrical Power Systems to Completely Replace Fossil Fuels, Geological Survey of Finland, Report 42/2021

Michaux SP (2022) Private communication

Porter Z, Habli I, McDermid JA (2022) A Principle-based Ethical Assurance Argument for AI and Autonomous Systems. arXiv preprint arXiv:2203.15370.

Rawls, J., (1951) Outline of a decision procedure for ethics. The Philosophical Review, 60(2):177-197

Rawls, J. (1971) A theory of justice. Harvard University Press

Raworth K (2017) Doughnut economics: seven ways to think like a 21st-century economist. Chelsea Green Publishing

UN (2022) Sustainable development goals, https://www.un.org/sustainabledevelopment/ (accessed 16th October 2022)

Van den Hoven, J. (1997) Computer ethics and moral methodology. Metaphilosophy, 28(3): 234-248.

Van de Poel, I, Zwart, S.D. (2010) Reflective equilibrium in R & D networks. Science, Technology, & Human Values, 35(2): 174-199

Wiedmann T, Lenzen M, Keyßer LT, Steinberger JK (2020) Scientists' warning on affluence. Nature communications;11(1):1-0.

Zemo (2021) Powered Light Vehicles Life Cycle Analysis Study, https://www.zemo.org.uk/news-events/news,powered-light-vehicles-can-enable-transport-decarbonisationlifecycle-analys_4329.htm (accessed 16th October 2022)

Introducing Autonomous Systems into Operations: How the SMS has to Change

John McDermid, Mike Parsons

Assuring Autonomy International Programme (AAIP)

University of York, UK

Abstract *When an autonomous system is deployed into a specific environment there may be new safety risks introduced. These could include risks due to staff interacting with the new system in unsafe ways (e.g. getting too close), risks to infrastructure (e.g. collisions with maintenance equipment), and also risks to the environment (e.g. due to increased traffic flows). Hence changes must be made to the local Safety Management System (SMS) governing how the system is deployed, operated, maintained and disposed of within its operating context. This includes how the operators, maintainers, emergency services and accident investigators have to work to new practices and develop new skills. They may also require new approaches, tools and techniques to do their jobs. It is also noted that many autonomous systems (for example aerial drones or self-driving shuttles) may come with a generic product-based safety justification, comprising a safety case and operational information (e.g. manuals) that may need tailoring or adapting to each deployment environment. This adaptation may be done, in part, via the SMS. This paper focusses on these deployment and adaptation issues, highlighting changes to working processes and practices.*

2 Introduction

1.1 Background & Rationale

Why a Safety Management System? Recent understanding of how accidents and incidents happen puts more emphasis on the causal factors external to the system and the organisational factors that contribute to errors being made (CAA, 2022).

The latter factors include how the organisation operates, how it sets out its procedures, how it trains its staff and what level of importance it gives to safety issues identified. A Safety Management System (SMS) addresses this and allows a proactive approach to safety by identifying causal factors and acting before an event happens. An SMS can therefore contribute to improving safety through a greater understanding of the hazards and risks affecting safety in the organisation.

In summary an SMS is the set of processes, procedures, management activities and cultural aspects that an organisation uses to ensure safety in its operation. Two useful definitions are:

> "... a systematic and proactive approach for managing safety risks...[an] SMS includes goal setting, planning, and measuring performance. An effective safety management system is woven into the fabric of an organisation. It becomes part of the culture; the way people do their jobs" (CAA, 2022), and:

> "Safety Management Systems for product/service providers ... integrate modern safety risk management and safety assurance concepts into repeatable, proactive systems. SMSs emphasize safety management as a fundamental business process to be considered in the same manner as other aspects of business management." (FAA, 2022).

These two SMS definitions include a common set of four process areas for components of an organisation:

1. Safety policy and objectives (management commitment, plans, methods, processes, and organizational structure needed to meet safety goals);
2. Safety risk management (new or revised risk controls based on risk identification and assessment of acceptable risk);
3. Safety assurance (evaluates the effectiveness of risk control strategies; supports the identification of new hazards);
4. Safety promotion (training, communication, and other actions to create a positive safety culture).

Explicit SMSs exist in domains other than aviation. The European Union Agency for Railways (EUAR, 2022) has some concise statements that help to frame the nature of a typical SMS:

> "The purpose of the SMS is to ensure that the organisation achieves its business objectives in a safe manner and complies with all of the safety obligations that apply to it...[it] enables the identification of hazards and the continuous management of risks related to an organisation's own activities, with the aim of preventing accidents...an SMS will provide an organisation with the necessary confidence that it controls and will continue to control all the risks associated with its activities, under all conditions...The SMS integrates into the business processes of the organisation...The SMS should be a living set of arrangements

which grows in maturity and develops as the organisation which it serves does so"

This paper focuses on changes or additions to the SMS where autonomous systems (AS) are introduced into an existing environment.

1.2 Context

We consider a new AS deployed into a specific environment. It could be an automated shuttle starting operations on a university campus; it could be an automated pallet system introduced into a factory; it could be a drone used by the military on the battlefield for the first time. It is assumed that this system will be delivered with a (product) safety case report, and that this comes with generic operational guidance and manuals covering a range of expected application situations. This guidance may reference a CONOPS (Concept of Operations) and also may have some operational restrictions or limitations which need to be observed. There may be a separate deployment safety case, demonstrating how the AS will meet safety claims for the specific environment, but this is unlikely for generic AS and is not assumed in the analysis presented here.

The AS is likely to be deployed in a staged process (see figure 1), involving[1]:

(i) Commissioning (pilot, trials, introduction into service),
(ii) Remote-controlled operations (possibly involving partial autonomy),
(iii) Fully autonomous operational service, and finally
(iv) Withdrawal from service.

In all these phases additional support is required via processes that include: (a) Monitoring, (b) Changes via maintenance and upgrades to functionality and (c) Incident and accident management. There may also be some generic supporting activities for the particular site or environment that the AS is working in; these include staff training and competency management.

---

[1] Stages could be based on the models of levels of autonomy used in, for example, the automotive sector. In this case, a vehicle may initially have an operator who can intervene while the vehicle drives itself before the later stages where there is no driver.

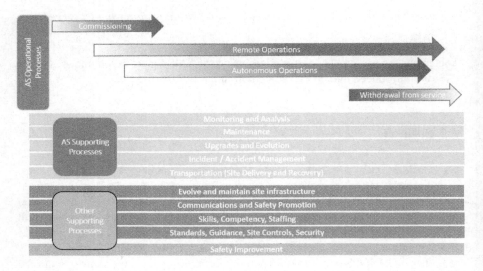

Fig. 1. Generic Deployment and SMS Areas for an Autonomous System

All these processes have to be updated for the introduction of an AS. For brevity, this paper focuses on five of these areas.

1.3 Structure of this paper

The introduction in section 1 is followed by a description of the work currently being undertaken in the AAIP to produce a framework for production of additions to an SMS to support AS (informally the "SMS delta"). Section 3 is the main part of the paper that considers selected SMS areas and discusses the nature and type of the additions (and perhaps changes) required when introducing an AS. Section 4 presents some conclusions on the work so far, and section 5 outlines some areas considered for future work.

2 SODA

The University of York AAIP programme (AAIP, 2022) is currently developing a management framework for the Safety of Deployed Autonomous Systems, (SODA), (SODA, 2023) as part of a family of developments for AS including assuring the machine learning (ML) based system components (AMLAS, 2022) and assurance of an AS within a complex environment (SACE, 2022).

SODA produces the AS-specific elements of the SMS for operation of the AS at a specific site. SODA is a process for systematic construction of an "SMS delta", i.e. the changes required to the SMS to enable safe operation of an AS. It assumes that there is already an SMS in place for operations at the site.

The result of applying SODA is a set of AS-specific processes and procedures to add into a standard SMS, including identifying a set of tangible inputs and outputs (documents and other artefacts).

SODA comprises a set of processes covering the activities identified in Figure 1. Each process comprises a set of process steps with inputs and outputs for each step. The process for the Commissioning activity from Figure 1 is shown in Figure 2. By working through the process steps, an addition to the SMS will be produced supporting the pre-operational commissioning of the AS at that site. The first three stages produce procedure fragments that are drawn together into a coherent procedure based on the supplied template.

The intent is that the SODA processes will be undertaken by safety professionals – site safety managers, safety engineers, etc. with the resultant processes and procedures produced in company-standard form to inform operators on the site or in a remote operations centre (ROC).

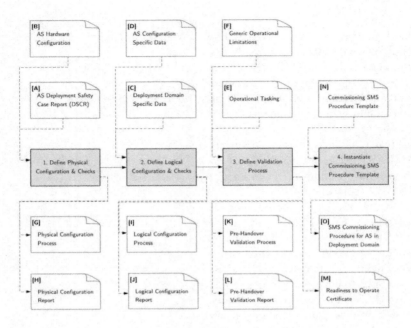

Fig. 2. Illustrative SODA Process for Commissioning

3 Selected Additions

This section considers some of the additions to the SMS needed for safe deployment of an AS, as might be expected to be derived from SODA. There could be many areas of the SMS that need updating. In this paper five of these areas are examined:

1. Remote Control and Autonomous Operations
2. Monitoring and Analysis
3. Upgrades and Evolution
4. Skills, Competency, Staffing
5. Incident / Accident Management

Figure 2 below shows how these areas relate to the original SMS scoping:

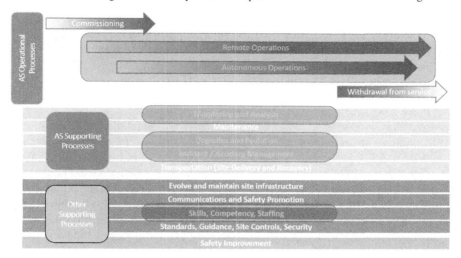

Fig. 3. Coverage of selected SMS areas from Figure 1 in this paper

So, in each of the four typical SMS areas we consider the following as needing review and/or update:

1. **Safety policy and objectives**. This needs to include contracts and agreements with the AS supplier and maintainer including Service-Level Agreements (SLA), that may also need to involve the site operator. These will include things like hours of support and call-out times. They will also include agreements - how much the AS supplier will attempt to do remotely and how often they may be required to attend site. Objectives should also be updated to cover the autonomous aspects, given that certain aspects may not be predictable such as object recognition via machine learning. Targets for safety performance of the AS in its context should be set. Policies that may need to be in place regarding the AS include communications recording and playback facilities, data storage and retention, data sharing protocols and engagement agreements with certification / accreditation bodies, regulators[2] and accident investigators. It is important to ensure that everything is in place so that any accidents or incidents involving the control of the AS can be thoroughly investigated using replayed data.

2. **Safety risk management**. This SMS element needs to consider issues of loss of communications, remote/local/autonomous behavioural conflicts, and interactions with personnel on site, plus other autonomous systems, etc. New pro-

---

[2] It is important to note that in certain sectors (e.g. nuclear, aviation) the regulator has an important role, and the agreements and methods of engagement put in place may need to be comprehensive. For instance, a regulator in these industries may want to be involved in reviewing any change applied to the AS, before it goes live.

cesses and protocols may need to be defined - on-site maintenance where remote operation must be disabled and autonomy isolated. New hazards relating to conflicts may be added by personnel who have to interact with the AS on a daily basis.
3. **Safety assurance**. The frequency and type of monitoring needs to be defined so that all issues can be recorded, analysed and eventually their resolutions audited. The effectiveness of the SMS on the organisational safety performance must be evaluated and reviewed. When issues are found there may need to be temporary changes to operations (e.g. a fault found with the autonomy means that the AS is disallowed in certain areas of the site). Clearly the existing Safety Case for the site(s) will need to be reviewed in the context of including the AS.
4. **Safety promotion**. Staff will need to be trained and regularly informed about how to interact with the AS, and what functionality is currently enabled. They will need a mental model of the AS behaviour that can be developed and enhanced through regular briefs. They need to be confident that the AS is fully under local control (i.e. autonomy isolated) when performing various duties, e.g. recharging and cleaning.

3.1 Generic Additions to the SMS

Generic additions to the SMS will be required taking into account the following (some of which can be seen as inputs to the Commissioning process in Figure 2):

Table 1. Generic Additions to the SMS

Name	Description
Operating Scenarios	Scenarios (typically defined as Use cases) that describe the expected behaviour patterns of the AS at the site.
Deployment Domain Specific Data	This may include maps of the deployment domain indicating key points, e.g. ingress and egress, no-go areas for the AS, etc. Local information may be needed, e.g. areas of the site prone to communications dropouts, the location of the "home" charging station, and "muster points" in the event of emergencies.
Generic Operational Limitations	Constraints on the operational design model (ODM) that will apply to the AS for an extended period (perhaps throughout its operation), e.g. temporal (can't operate outside working hours), or geographical (terrain where the AS is likely to get stuck, and thus must avoid), etc.
Operational Tasking	The task or set of tasks that the AS is expected to do, e.g. traverse the site looking for particular features or performing maintenance actions.

Name	Description
Hazardous Scenarios	A subset of the scenarios that are identified as hazardous; note that these will likely be included in the Deployment Safety Case report (if there is one).
Existing Site SMS / Safety Procedures for Site	The pre-existing safety management system for the site(s). This may consist of higher-level documents that flow down to all sites. This may be specific to the contractor(s) that operate and maintain the site.

3.2 Remote Control and Autonomous Operations

When an AS is deployed it is often able to be remotely controlled as well as capable of fully autonomous operation[3]. However, this ability creates new risks and complexities as these modes of operation may conflict. An example might be if the AS is a vehicle, and the remote operator mistakenly drives the AS at another vehicle or a static object potentially causing a collision. In this case the AS may try to avoid the collision by disobeying the remote-control commands.

Hence, there may need to be operational rules (protocols, procedures, check lists) added to the SMS to ensure that a remote operator is fully aware of the situation the AS is in before issuing commands.

In reality, the interaction between the operator and the AS is likely more complex than this: the AS may have layered levels of functionality (including avoidance of harm behaviours and self-protection), and the remote control may only be able to override some of the functionality, even if trying to avoid an accident.

Who has control of the AS and who is responsible when accidents occur could be a very difficult problem with typical remote, local and autonomous control: there may be several levels of remote operation, e.g. via an operations centre and via a local hand-held controller. Conflicts with both these and the autonomous functions need to be managed through SMS procedures and protocols, at least in part.

In this context, the SMS will need to take account of:

---

[3] In some industries (e.g. UK civil aviation) autonomous operation is defined as **only** where there is no possibility of human intervention. If there is a remote link in place and a remote operator can intervene (to some extent) then it is not autonomous but rather "automatic with high authority", and this is an important distinction. In some safety-critical industries, SMSs have been developed over many years to deal with those types of system and interactions involved. The real issue is where intervention is not (realistically) possible, e.g. some vehicle scenarios (land/air) or a cobot that works in a factory in an uncaged environment alongside humans. Also where AI is used to support the AS decision making; here SMSs need to be updated to account for non-deterministic (or at least not easily explainable) behaviours and decisions made and their effect on staff.

Table 2. Remote Control and Autonomous Operations

Name	Description
AS Message Flow Definition	Messages sent to and received from the AS either from a remote operations centre (ROC) or a hand-held device.
Primitive Procedures Definition	The set of basic procedures for interacting with the AS, e.g. to isolate autonomy (and report that this has been done), to start a pre-stored task, to report completion of the task, to "request assistance", etc.
Remote Operations Centre Procedures	Processes and procedures for the operations staff to follow at the ROC (may cover many ASs and many sites).
Local Operating Protocols	Step-by-step instructions as to how to manage local operation and autonomous operation of the AS at site

3.3 Monitoring and Analysis

In conjunction with the local / remote control issue, there is likely to be a need to have real-time or near real-time monitoring of an AS, to ensure everything is working within safe bounds. This is most likely done using radio networks that are subject to drop-outs, delays and interference (both unintended and intentional). Hence processes and procedures in the SMS are required to establish what to do if communication is lost. Typically, a certain level of communication loss is tolerated but after a period must be re-established or the AS will have to perform contingency actions, depending on the circumstances (e.g. execute some sort of minimum risk manoeuvre or abort its mission)[4] and enter a state enabling it to be recovered.

Analysis is required of the safety performance of the AS. This could be near real-time (e.g. via a safety dashboard) or slower-time analysis (each day, week or month). This requires that data is available from the AS, so must either be transmitted or stored for later uploading. Potentially, the amount of data produced will be very large, and it will be a significant effort to process it to look for early signs of faults that might lead to a safety event, or indeed no-fault cases that might cause a problem[5]. Processes and procedures for collecting, storing, processing and analysing the data from the AS will therefore be required in the SMS. These will likely have to involve third-party organisations to provide communications services, plus possibly cloud storage provided by web service providers.

---

[4] It is recognised that for aerial AS it may be safer to continue to an intermediate safe location or indeed to complete the mission if it is close to a landing zone or airport

[5] In this case although an individual or fleet of AS are working to their specification, the combination of behaviours (perhaps involving other manufacturers' AS or interactions with humans) is leading to unsafe situations

Table 3. Monitoring and Analysis

Name	Description
AS Monitoring Data	Data from interoception, assessing the state of the hardware, e.g. from actuator built-in tests (BIT), and from assessment of the impact of the environment on the sensing suite, e.g. impairment due to fog or rain. This might inform temporary operational limitations.
Environmental Monitoring Data	Information about the state of the deployment including from sensors within the infrastructure and potentially from the AS itself or other AS on site.
AS Safety Performance Analysis	A report produced regularly that demonstrates the required safety performance is being met. Compilation of this report may require data from the manufacturer as well as from site.
Agreements with Communications Providers	Increased site communications needs may require changed agreements with communications providers, covering specific service levels including assurance and integrity targets
Agreements with Cloud Storage Providers	There will be a large amount of data produced by the AS and associated infrastructure and this will need to be stored in a secure cloud environment.
Data Sharing and Retention Policy	It is critical that data that has been saved (either stored in the AS or via site infrastructure) and is not lost (either deleted or overwritten). It must also be able to be shared with the appropriate parties: manufacturer, maintainer, site owner, independent investigator, etc[6].

3.4 Upgrades and Evolution

Changes to an AS will typically be made via Over-The-Air (OTA) updates that can be done on site or in the field wherever communications are possible. OTA updates will be governed by the SMS (for instance, there may be restrictions on when an update can become active, and where it can be trialled) that may affect both the operating software and data used within the AS and site infrastructure. The data used by the AS may be of several different types, including:

1. Configuration Data (to configure features within the AS itself)
2. Navigation Data (including maps, allowable routes, prohibited areas, etc)
3. Site Data (including changes to site infrastructure locations, etc)

---

[6] It is recognised that some data may need to be post-processed (e.g. anonymised) before sharing, for example, to obfuscate faces of people that a camera may have captured. GDPR requirements may apply.

4. Machine Learning Training Data, that may influence behaviours such as navigation or object recognition

What is different here is that the software and data will largely govern the behaviour of the AS, and that the changes in behaviour need to be understood by people working on site.

Of course, changes may also be introduced to the hardware of the AS. Sensors could be replaced, functionality could be upgraded (e.g. higher-capacity batteries) and new features added (e.g. additional cameras). Changes may be undertaken at site or the AS may have to be returned to the manufacturer for the change to be performed. Different SMS processes and procedures are required in each case, for instance, AS removal from site and AS delivery to site. In all cases the disabling of autonomous functions, and verification of this action is paramount and procedures for this will have to be built into all SMS processes and supported by the AS itself. There may also need to be a proving area or testing ground at site where changes can be tested and verified, again this would largely come under site procedures within the SMS[7].

Table 4. Upgrades and Evolution

Name	Description
Change Management Procedures	The various procedures and processes to be used to make changes to the AS and the supporting infrastructure. May involve change at different locations and different types of change (hardware, software, data).
OTA Update Protocol	The steps to enable safe changes via OTA updates. Should cover fleet upgrades, mixed fleet issues and backing out unwanted changes.
Upgrade at Site Procedures	Procedures for making change at site, including isolation of autonomy.
Testing Ground Definition	If the changes are to be tested at site then a definition of the test ground (environment) will be needed.
Testing Ground Procedures	If the changes are to be tested at site, then detailed procedures will need to be established for testing in a safe manner, away from operational infrastructure.

3.5 Skills, Competency, Staffing

The SMS will have to cover areas of staff training and competency for dealing with the AS. Depending on the nature of AS, this could range from simple awareness

---

[7] It is recognised that there may be local site operators operating under service contracts with their own SMS in place. In this case the prime should ensure that any higher-level SMS requirements related to the AS flow down to local operators as needed.

courses (for an AS unlikely to cause any harm) to specific and detailed training with examination and certification for larger, faster or perhaps airborne AS. This training will likely have to be tailored for the specific site to include local conditions (for instance, including procedures to deal with the muddy conditions or flooding for a land vehicle AS). The training should be such that it enables staff that may come into contact with the AS to interact with it safely and to minimise any operational difficulties for them or the AS. The SMS will have to contain mechanisms for staff to report issues with the AS, and to be able to be informed of updates and changes in behaviour as a result of reports.

Of course, staff do not always behave as they should, and it is possible that protective actions are forgotten or ignored (e.g. not isolating autonomy before maintenance actions). Also, unauthorised actions or dangerous interactions with the AS may take place (for instance, 'playing chicken' in front of a vehicle AS), putting people at risk. In this case the SMS needs to anticipate as much foreseeable misuse as possible and contain warnings in manuals and provide regular training covering misuse. It may have links to site disciplinary procedures to deal appropriately with any actual misuse to discourage recurrence.

Table 5. Staff Training & Safety Information

Name	Description
Staff Handbook	The staff handbook should outline how staff are expected to work with the AS, and detail warnings and limitations.
Staff Training Courses	Staff may require training before being allowed near the AS.
Staff Briefings	Staff may require regular briefings if the AS functionality changes due to frequent OTA updates.
Incident and Fault Reporting	There may have to be changes to the standard site incident reporting procedures (e.g. additional statutory information required) when logging an incident involving an AS.
Staff Welfare and Support	Existing site staff may well feel threatened by the AS if it performs duties previously done by them. They may need retraining, redeployment and support services to manage the safe introduction of the AS.

3.6 Incident / Accident Management

An operational AS will require processes and procedures for managing accidents and incidents; dealing with them is an important part of any SMS. This will include everything from managing communications with the press and accident investigators, through analysis of the causes, to making the accident site safe.

Hence, with an AS, the parts of the SMS that require modification include policies for communications (and site security[8]), noting that there may be a lot of press interest in a major accident involving an AS, changes to procedures such as those used for accident management including how to establish a safe site, instructions on how to verify that the autonomous functionality is disabled (and cannot be mistakenly re-enabled remotely), and how to recover the AS. There may also be additional processes for how to deal with investigations internal to the organisation and also with external accident investigators. If people are involved in the accident there may be injuries to deal with and emergency services may need to operate special protocols when autonomy is involved, requiring independent verification of autonomy isolation to ensure that emergency services staff are not put at undue risk.

Hence the existing SMS may need to have updates related to:

Table 6. Accidents & Investigations

Name	Description
Agreements with Manufacturers and Maintainers	It is important that the manufacturer of the AS and any maintenance organisation are 'on side' and able and willing share information and assist with any incident recovery and analysis.
Agreements with Independent Investigators	Agreements, permissions and working methods need to be in place with any independent investigators in advance.
AS Manufacturer Supporting Information	Any additional information that the manufacturer of the AS has regarding management of incidents or accidents, e.g. autonomy isolation, towing considerations for land-based AS, etc.
Site Information for Accidents	Any site-specific information regarding accident management, e.g. site procedures for turning off power; isolation of autonomy; chain of command; emergency services call-outs; fire routes, etc.
Tools and Equipment Required	The AS may require special handling, tools and equipment (e.g. for recovery and towing).
Skills and Competencies Required	The training, skills and competencies staff must have in order to deal with an accident involving the AS.
Incident Handling Process Definition	A site-specific process defining the staff, procedures and actions relating to handling a site incident involving the AS. May involve everything from handling the press to isolating part of the site.

---

[8] It is recognised that the inter-relationship between an SMS and security management processes is an important one. There can be issues and conflicts to resolve (for instance in sharing of operational data about the AS and access to certain areas of a site for accident investigation). This will be the subject of future work.

Name	Description
Configuration, Software and Data State of the AS	It is important that the configuration data and software state of the AS is preserved after an incident so that it can be analysed[9]. This is likely to require a detailed procedure on site, especially if communications with the AS is lost.
Configuration, Software and Data State of Site Infrastructure	It is important that the configuration data and software state of the site infrastructure (including other AS) is preserved after an incident so that it can be analysed. This is likely to require a detailed procedure on site.
Temporary Operational Limitations	Constraints that apply for a limited period of time, e.g. spatial limitations whilst recovery operations are carried out, or temporal limitations due to what is happening on site.

4 Conclusions and Discussion

This work has examined changes to an existing SMS when an AS is introduced. Five areas were discussed, and changes proposed. The nature, scope and scale of changes will depend on the deployment context (e.g. land, air, water, space) and characteristics of the AS (e.g. size, weight, actuators, proximity to staff). However, the generic AAIP SODA framework is designed to be comprehensive and detailed enough to be applicable to a wide range of AS modalities. Whilst not discussed in detail above, it is expected that developing the SMS will identify derived requirements on the AS, e.g. to be able to report that autonomy is isolated, to move to a "muster point" if informed of a site emergency. Thus, a level of co-design between the AS and the SMS may be valuable

It is acknowledged that an SMS is rarely developed from scratch, instead it is a combination of knowledge, process, procedures and instructions that evolves over time and generally develops incrementally. Therefore, it has been assumed that only changes (deltas) need to be made to cover the AS introduction. However, it may not be so simple, as there may be conflicts with existing processes and procedures, and these may also need to be updated. Lastly, it may not be obvious what should be done in cases not covered by the SMS (perhaps collisions with other diverse AS that should not be on site, or unexpected human behaviours when interacting with the AS). Therefore, it is recommended that a full review of the SMS is undertaken after introducing the changes for the introduction of an AS.

---

[9] Note that local storage of data (vehicle, telemetry) for accident investigation purposes may impose design requirements on the AS, necessitating a comprehensive "black box" in effect; the SMS may also need to support data transfer and storage as well. This is the subject of future work.

5 Further Considerations

This section outlines some additional work that could be undertaken to further elaborate what is required for an SMS dealing with operations incorporating an AS.

Firstly, the other identified areas of the SMS require analysis to see what changes may be required. In addition, the following areas are likely to be important considerations when developing the SMS:

Regulators: In many safety-related industries (at least in the UK) there is strong regulation in place, backed up by legislation. This applies particularly to high-risk and well-established sectors such as nuclear, civil aviation and rail. It is expected that the regulators in these industries will take an active role in the introduction of an AS, setting out objectives and requirements, producing guidance, reviewing and approving changes to an AS, monitoring operational safety performance, and making recommendations for safety improvement. Regulators may want to be involved in monitoring other aspects such as staffing issues (effects on absences, rostering, etc.) and the impact of introducing AS on safety culture within an organisation.

Digital Twins: When an AS is deployed into an operational environment there may be a need to integrate it into the maintenance and operational models that the organisation uses to monitor and maintain the operational status. For complex sites the models utilised can be very sophisticated. For this purpose, it may be necessary to create a digital twin model of the AS in its environment. This model can be as abstract or detailed as required but may have to include modelling of autonomous decision making[10]. As an example, if the autonomous decision making within a land vehicle (say farm vehicle) always chooses a particular route over rough ground, this ground may become muddy and impassable and therefore increase risk over time.

Replacement of the AS: When an AS is upgraded for a newer model or different variant, the SMS will need to be reviewed to see if any further changes are required. In fleet situations, the picture is more complex as generally not all AS can be replaced in one go and operation with a mixed fleet of, say, older and newer models together may be more likely. In this case the risks of operating with different AS models or versions together must be assessed and mitigated.

Removal of the AS: Where an AS performs a safety function, e.g. fire detection or suppression, then it cannot be removed without an increase in risk. This risk will have to be managed, either manually (e.g. increased monitoring by staff), or by replacing the AS by other systems (that may or may not have autonomous functions).

---

[10] Some digital twin models can take data from the physical system to continually refine the digital model.

Phased Introduction: It is recognised that an AS is usually not introduced into an operational setting in one go; typically, there may be a series of iterative trials, phases or stages where more functionality is exercised and additional parts of the domain (say areas of the land site or air space) are included in each phase. In this case, full safety documentation is unlikely to be available at the start of trials, so it is important that appropriate mitigations are put in place to address the risks at each stage, noting that these might be progressively reduced as confidence is built in the safe operation of the AS.

Acknowledgments We thank Paul Hampton for very valuable review comments.

Disclaimers All views are those of the authors and not their respective organisations.

References

AAIP (2022), Assuring Autonomy International Programme, https://www.york.ac.uk/assuring-autonomy/ , accessed October 2022

AMLAS (2022), Assurance of Machine Learning for use in Autonomous Systems (AMLAS), AAIP, https://www.york.ac.uk/assuring-autonomy/guidance/amlas/ accessed November 2022

CAA (2022), Safety Management Systems: Guidance for small, non-complex organisations, CAP 1059, https://publicapps.caa.co.uk/docs/33/CAP%201059%20SMS%20for%20small%20organisations%20(p).pdf , accessed October 2022.

CAA (2022a), Safety Management Systems (SMS): guidance for organisations, CAP 795 https://publicapps.caa.co.uk/docs/33/CAP795_SMS_guidance_to_organisations.pdf, accessed October 2022

EUAR (2022), ERA->Activities->Safety Management System, https://www.era.europa.eu/activities/safety-management-system_en , accessed October 2022

FAA (2022), SMS Explained, https://www.faa.gov/about/initiatives/sms/explained , accessed October 2022

SACE (2022), Guidance on the Safety Assurance of autonomous systems in Complex Environments (SACE), https://www.york.ac.uk/assuring-autonomy/guidance/sace/ , accessed November 2022

SODA (2023), Guidance for the Safety of Deployed Autonomous Systems (SODA), AAIP University of York, in preparation

The Language of Risks and the Risks of Language

Catherine Menon[1], Austen Rainer[2] and Lorna Gibb[3]

[1]University of Hertfordshire

[2]Queens University

[3]University of Stirling

Abstract *The language we use when discussing risk plays a major role in how that risk is perceived. In this paper we present a discussion of risk-based language, illustrating how the word, sentence and structure choices we make can serve to obscure or exaggerate certain risks. We introduce the concept of a system's narrative, which describes the context and history of this system, and explore how different narrative techniques can provide different perspectives on the risk posed by the system. Finally, we discuss metaphor, simile and language creativity, and whether there is a place for these in safety-critical analysis.*

© Catherine Menon, Austen Rainer and Lorna Gibb 2023.
Published by the Safety-Critical Systems Club. All Rights Reserved

Rising to the challenge of certifying automated vehicles

Jamie McFadden

Vehicle Certification Agency

Abstract *Automated Lane Keeping Systems (ALKS) are the first commercially available systems designed for passenger vehicles that will enable the driver to safely hand over control to the vehicle. This is made possible through certification to UNECE Regulation 157, the first Type Approval Regulation for automated vehicles. Combined with the adoption of UNECE Regulation 155 on Cyber Security the first steps have been taken towards safe and secure deployment of automated vehicles. Looking at these two topics we can see how vehicle regulators are rising to the challenge of certifying automated vehicles that will enable widescale commercial deployment.*

© Jamie McFadden 2023.
Published by the Safety-Critical Systems Club. All Rights Reserved

Hierarchical Approaches to Product Cyber Security: An Automotive Case Study

Robert Oates, Aditya Deshpande

BlackBerry Ltd.

UK

Abstract *Securing the operation of an out-of-context component can be extremely challenging. This is due to a number of reasons, not least that the context of use for the system component has a huge impact on the exposure of the system to specific risks. Many standards across multiple sectors focus on the role of system integrators and Tier 1 suppliers. But how should that security argument flow down to Tier 2 suppliers and below? Can Tier 2 suppliers be "intelligent suppliers", providing security assurances that feed into hierarchical or modular assurance cases? In this paper we approach these questions, illustrating the challenges and proposed solutions using the use case of a safety-assured, automotive operating system. The automotive sector was selected because the cybersecurity standard (ISO 21434) demands that the security argument extend beyond safety-related security issues. An operating system is a highly versatile component that can have multiple contexts. The paper concludes that there are activities that lower-tier suppliers can do to support the integrated security/safety argument. These activities are then highlighted as potential requirements for system integrators, Tier 2 suppliers and below.*

1 Introduction

Supply chains within the automotive industry are commonly divided into tiers, where Tier 0 is the overall systems integrator, Tier 1 is a high-level component integrator, and Tier 2 is a low-level component provider, see Fig. 1. Frequently these tiers need to be extended beyond 2, especially for software, where layers of software are built on other libraries further down the supply chain.

© Blackberry Ltd. 2023.
Published by the Safety-Critical Systems Club. All Rights Reserved

In an ideal world, security concepts are developed at the top tier and flowed down to subsequent levels. This ensures that there is consistency of approach to security for the end-product and that there are no security holes generated by "gaps of uncertainty" between suppliers. However, security concerns about flowing information about vulnerabilities to lower tiers, and intellectual property fears can prevent this process being as transparent as necessary to ensure a secure and safe system. In this paper we explore what lower-tier suppliers (Tiers 2 and below) can do independently to support the security (and as a result, the safety) of the overall system. Chapter 2 will explore the challenges of securing out of context components, Chapter 3 will provide solutions to some of those challenges and Chapter 4 will present a case study illustrating these points, using an automotive operating system as an example.

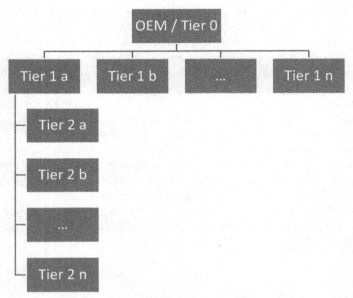

Fig. 1. Tier Structure for Automotive Manufacturing

2 Challenges for securing out of context components

The core practical challenge for a lower-tier supplier stems from the propagation of security risk and security strategy down the supply chain from higher tiers. Frequently system integrators are understandably nervous about sharing their security strategy to lower tiers, as it could be useful information for attackers, especially if that information contains architectural or vulnerability information.

In the absence of context Tier 2 suppliers have the option of producing bespoke products for each project or producing a general-purpose product that can be applied in multiple contexts. The former option, despite techniques such as product lines, is much more costly. The latter option presents the problem of generating a system that can be safe and secure across a number of use cases, with little prior knowledge of the contexts it will be used in.

A general-purpose product must contain a multitude of features, in order to provide the contextual flexibility required, this is an anathema to safety and security pressures to keep systems as simple as possible and makes exhaustive testing impossible. In addition, where features are security controls, a lower-tier supplier has little to no say over which of those controls are enabled in the final system. If the product is then involved in a malicious attack, the supplier may find their own reputation being tarnished through no fault of their own.

At a general level, the lack of context means that risk management is extremely challenging. The impacts of a risk manifesting are entirely context driven, and the likelihood is a function of the impact (with certain impacts being more desirable for an attacker than others) and the system architecture (which controls the accessibility the attacker has to the target system).

3 Solutions to the challenges

The authors of this paper posit that challenges presented by securing an out of context system can be addressed, in part, by three key techniques: 1) Flexibility (of features and tooling), 2) Secure development processes, 3) Compliance support.

3.1 Flexibility

For a lower-tier supplier, flexibility is key. By providing a highly configurable system, it is possible to adapt to top-down security concepts. Common design wisdom is that features which support both the safety and security of the product should be enabled by default. However, this is not always practical. For example, if encrypted communications is a default setting, but not used by the integrator, the Tier 2 product will not work "straight out of the box", which adds developer friction, and could make the product less competitive.

3.2 Secure Development Processes

One pathway for a lower-tier supplier to weaken the security case of the overall system is to introduce new vulnerabilities and backdoors which are available from the integrated system. Adopting a secure development process at either the system level (Ross et al. 2016) or software level (Souppaya et al. 2022) during the creation of the lower-tier product reduces the number of vulnerabilities.

A core concept for secure development, is the integration of a risk management system into the development process. This re-introduces the challenges of generating a meaningful risk assessment in the absence of context. For a lower-tier system the best strategy for overcoming these issues is to generate a portfolio of use cases based on prior applications of the product.

3.3 Compliance Support

It is not always possible to assure an out-of-context component to any particular standard. A notable exception being Common Criteria (ISO/IEC 15408-1 2022), which allows, specifically, for a *security* component to be tested in isolation. However, it is possible for a component to be developed that supports a wider risk argument (whether that be safety or security), making compliance easier for the system integrator.

The concept of modular security and safety cases has been posited before (Kelly 2003). In order to prevent an explosion in the argument complexity, suppliers can ensure that vulnerabilities inherited from further down the supply chain are addressed at their level, without the integrator needing to follow the supply chain thread any deeper. This requires the supplier to have a comprehensive understanding of the software that they are building upon, and to make the argument that the risks are addressed, either through patching or architectural means. At the core of such an argument is the SBOM (Software Bill of Materials) which explicitly lists the libraries being used.

Of course, historical use of the component may inform future use, so suppliers can attempt to add context, and de-risk future endeavours, through the application of risk management techniques on a use-case portfolio. However, in-depth techniques that are not scalable should be avoided to prevent rework. Two techniques to be considered are the TARA (Threat Assessment and Remediation Analysis) (Wynn 2014) and CRAF (Cyber Risk Assessment Framework) (McDermid et al. 2018). For a TARA, the first step is the production of a system model, with clearly defined boundaries. Such models can be highly context-dependent, but if the component is part of a product line, it may be possible to reuse analysis for specific interfaces. For a CRAF the first step is to identify crucial data properties that are required for the safe and effective operation of the system, which then inform the security properties that need to be protected to ensure the same goals. Whilst context may change the properties required, for a common component there is likely to be a significant overlap between use cases.

4 An Automotive Tier 2 Product Case Study: The QNX Operating System

The QNX embedded operating system (QNX 2022) consists of a microkernel and a set of processes (file system managers, network managers, device managers, process managers, etc.) providing the open systems POSIX API (IEEE 1003.1, 2017) in a reliable, robust, and scalable form suitable for a wide range of real-time and distributed domains (e.g., automotive, medical, rail). The micro kernel provides fundamental services such as synchronisation, message parsing, scheduling, etc. to be utilised by the system and user processes. The name 'microkernel' is because the critical OS components such as device drivers, networking stacks, and GUI components are treated as individual processes with their own protected memory partitions and the kernel provides a software bus in the form of inter-process communication for them to communicate. The flexibility of the OS means that it is applied in multiple industries and multiple applications within those industries.

4.1 Flexibility

The QNX operating system has a number of selectable security features that can be configured to operate within a number of potential use cases.

The ability to set security policies to define and govern what individual processes are allowed to do gives system designers and integrators the ability to manage security in a centralized way.

Other inbuilt software security features include trusted execution to provide boot time integrity, adaptive partitioning to protect against denial-of-service attacks, file system integrity, file system encryption, access control, network security.

Selecting which security features to enable by default, and which security features should be enabled through configuration is a non-trivial task. The two most prominent default functions are the use of address space randomization to prevent attackers from being able to reliably execute malicious code through memory traversal, and the separation of functions into separate memory addresses to prevent malicious code propagating through the system.

The features which are not enabled by default are typically selected based on the reduction of "developer friction", i.e. features that would make software development more complex if introduced too early in the development process.

To prevent additional developer friction, the QNX SDP (software development platform) provides standardized components for connectivity, media management, and screen management to prevent the introduction of implementation-specific vulnerabilities.

As a software development company, QNX does not provide any debugging tools such as on-board diagnostics or a reprogramming tool. Although the software does include debugging clients for the use of system integrators to test and validate their end system. The responsibility of deactivating these debugging tools in the production environment lies with the system integrator. Hence, it is vital to inform the customer regarding this before going into production. In addition, software development tool chains, compilers etc. should also be verified for vulnerable components. The communication of such forward-flowing requirements is left to the production of a security manual (similar to the safety manual) that clearly outlines integrator responsibilities.

4.2 Secure Development

Traditionally security has been the remit of corporate IT Security, who in themselves are systems integrators and not directly software developers. This means that a major step towards securing the product requires the development of:

- A security lifecycle policy describing the management and technical activities such as planning (security requirements, verification assessment, and audits), responsibilities, and description of the organisational view of the security management system;
- A security development lifecycle describing various standards used for security in the software development (e.g. secure coding guidelines), procedures for review of code for vulnerability fixes, testing procedures etc.

Blackberry/QNX currently has an ongoing project to enhance its secure software development lifecycle and works with its customers to develop work products to strengthen its secure development.

The introduction of security-focussed reviews during key development phases provides significant benefit for a lower-tier system as they require little contextual information to be effective. Review gates are placed at requirements, architecture, design, and implementation delivery points. Requirements reviews allow identification of product requirements impacting security. Architectural reviews provide a way to ensure security best practices are followed during the architectural definition (e.g., segregation of components, definition of security controls, introducing and utilizing security by design principles). Design reviews can highlight trade-offs between use of security protocols and performance. Implementation reviews can be utilized to verify that the security requirements are met, no new vulnerabilities are introduced, and best practices such as secure coding guidelines are followed.

Security activities during the verification and validation phase of the development lifecycle provide a way to understand the residual risk in the developed product. A variety of technological solutions such as static and dynamic analysis for security testing can provide Key Performance Indicators (KPIs) and the security gaps that were not identified during the development reviews (in addition to metrics on achieved security requirements). For components where third-party code is integrated into the product, binary analysis (tooling which can decompile an executable and identify subcomponents and libraries used) can provide useful insight in the form of software composition analysis, use of Open-Source Software (OSS), Common Vulnerability and Exposure (CVE) information and violations of coding standards.

4.3 Compliance Support

ISO 21434 Road Vehicles – Cybersecurity engineering is a standard that focuses on addressing cybersecurity engineering aspects of vehicles, products, and components in the automotive supply chain. Unlike the automotive functional safety standard ISO 26262, ISO 21434 does not prescribe specific ways of developing compliance and gives flexibility to the organisations in developing concepts and solutions to incorporate cybersecurity in the product's lifecycle. BlackBerry/QNX being a major tier 2 supplier for software in the automotive space is required to develop compliance and supports its customer including OEMs (Original Equipment Manufacturers) by providing various artifacts towards ISO 21434 and it's overarching legislation UNECE WP 29 (UNECE WP29 2022). Whilst the flexibility of implementation means that adapting ISO 21434 to lower tiers is possible, it is important to not only stick to the letter of the standard, but also actually support risk management of the overall system.

Organisational Cyber Security Management

The nature of cybersecurity incidents has proved that vulnerabilities can be introduced in a product through ways an organisation conducts its day-to-day operations. This has also been seen in functional safety where lack of safety management systems during development, operation and maintenance can lead to unsafe behaviour of the system (see (CAA 2013) for a wider discussion on safety management systems). As a result, having a cybersecurity management system using a risk-based approach, defining organisational policies and processes for cybersecurity, and developing a cybersecurity culture within the organisation and the supply chain is vital in developing secure systems.

From a BlackBerry/QNX perspective cybersecurity policies, rules and procedures defined to manage cybersecurity risk had been focussed on the corporate IT infrastructure. The policies were developed in accordance to ISO 27001 and vehicle components such as the QNX operating system remained beyond the scope. The risk management methodology was limited to identifying IT assets and focussed on servers, storage, and computing platforms. The organisation's incident response capability however, included identifying vulnerabilities in its products. Information sharing related to the products' vulnerabilities and security related information follow the corporate standards for sharing security related information.

As can be seen by the mix of IT and product policies, prior to the implementation of ISO 21434, managing product cybersecurity related process was conducted on an informal basis. After the initial implementation of ISO 21434, a key aspect of cybersecurity management is having defined roles at the leadership, process, and the product level to ensure all security activities are budgeted, standardised, and followed. Development of roles such as product security officer alongside the information security officer, project security managers, and security process owners along with security developers, testers, and engineers are vital for ensuring security activities are achieved.

In addition to process, it is vital for an organisation to have a well-managed cybersecurity culture. This implies that the organisation is investing in creating a competency for cybersecurity. As an organisation, BlackBerry conducted training and awareness programmes which were more focussed on IT and corporate security. From an ISO 21434 perspective, training programmes related to secure SDLC (See section 3.2 for references to system and software level SDLCs), TARA, security testing etc are valuable in developing competency. However, there is a lot that can still be achieved in terms of periodically evaluating training and awareness and ensuring that continuous improvement is applied from a security perspective. Lessons are learned from Safety where QNX safety-certified operating system conducted various improvements in the product based on safety analysis over a period.

Cybersecurity related incidents and vulnerabilities require an organisation to share critical information regarding the products and services. Having corporate standards that describe the type of information that can be shared, an approval process for sharing and redacting information, and vulnerability disclosure policies are vital. As a Tier 2 supplier, BlackBerry has taken steps to ensure that cybersecurity corporate standards for information sharing are also applied to its products in the automotive domain. This is vital to balance the need to communicate with external parties with the prevention of disclosure of confidential information and intellectual property.

Software defects can manifest as vulnerabilities, especially if those defects are in a security feature. For example, defects or issues in encrypting and decrypting data within a file system encryption function could cause information to be made unavailable or disclosed to a third-party. As a result, traditional software quality processes need to be employed to minimise the accidental inclusion of defects. Security issues can also arise in the software due to project scope change, feature change etc. Hence, having a change management system that considers the modification lifecycle, software defect management and impact analysis is vital from a security standpoint. Documentation management for software, e.g. software bill of materials, records of software versions, list of CVEs in the existing software are also important from a security point of view. Configuration management and source code control is vital from the point of view of bug and vulnerability fixes in ensuring that the bugs are fixed, and additional vulnerabilities are not introduced.

Information security management of the tool chains for software development environments e.g., JIRA, SVN, etc. are an important aspect of a cybersecurity management system. There is a possibility that a disgruntled employee can use these tool chains for malicious intentions causing inherent risk in the product. Hence, having access controls and permissions to these development environments is critical. In addition, reporting of access violations and management through a Secure Operations Centre is also critical.

Audits form an important aspect of maintaining cybersecurity and evaluating the maturity of the organisation's processes. Independent audits performed periodically can provide vital insights into areas of the organisation where the maturity needs to be improved. Audits of existing IT security management systems for ISO 27000 series, quality management for ISO 9001, Automotive SPICE (VDA QMC 2015), etc. provide vital supporting information for showing compliance to ISO 21434. However, an audit and assessment of specific projects before production release also provides vital evidence for organisational maturity and making sure that no risks arising from immature processes manifest as problems in the product. BlackBerry/QNX has a strong reputation for security and uses its internal teams to conduct audits and are currently expanding to cover the additional scope of ISO 21434.

SBOMs

The QNX operating system uses the BlackBerry Jarvis (Jarvis 2022) tool to decompile and analyse the binaries produced by the build process. Analysing the binary means that a validation exercise can be performed to ensure that the intended SBOM (i.e. the libraries that are deliberately included in the build) matches the actual SBOM (i.e. the libraries that are inherited through compiler linkage). In addition, Jarvis scans for data exfiltration (movement of potentially confidential information beyond the perimeter of the network it is stored on) by looking for embedded strings and ensures that there is a complete list of CVEs associated with each used library.

Modular Security Cases

All of the techniques discussed allow the case to be made that the component is adequately secure within an abstract context by addressing the key sources of vulnerability (inheritance of vulnerabilities, and implementation-specific vulnerabilities). However, as with component-based breakdowns of security cases, it is *not* possible to make a coherent argument that there will be no vulnerabilities as a result of interactions between system components at higher levels. For example, denial of service attacks generated by escalating comms between two components. Such dangerous interactions must be identified and eliminated at the higher level.

5 Conclusions

Lower-tier suppliers can perform activities that will feed into a greater cybersecurity case. These activities fall into the categories of: flexibility (to adapt their product into multiple security cases); secure development (to limit the introduction of new vulnerabilities); and compliance support (to assist the development of a coherent security case). The majority of the barriers to more in-depth support are related to the lack of context that a lower-tier supplier has for the specific use case of a system integrator. However, keeping a portfolio of historic use cases can inform future use cases, or at the very least prevent rework for similar uses.

References

CAA (2013) CAP 1059 Safety Management Systems: Guidance for small, non-complex organisations.
IEEE Standard 1003.1-2017 (2017) POSIX Operating System Standard
ISO/IEC 15408-1 (2022) Evaluation Criteria for IT Security Edition 4

Jarvis (2022) BlackBerry Jarvis https://blackberry.qnx.com/en/products/security/blackberry-jarvis Accessed on 25 November 2022.

Kelly, T.P. (2003) Managing Complex Safety Cases. Current Issues in Safety Critical Systems pp 99-115

McDermid, J.A., Asplund, F., Oates, R., Roberts, J. (2018) Rapid Integration of CPS Security and Safety. IEEE Embedded Systems Letters, Volume 11, Issue 4, Pages 111-114

QNX (2022) BlackBerry QNX https://blackberry.qnx.com/en Accessed on 25 November 2022

Ross, R., McEvilley, M., and Oren, J. (2016) NIST SP800-160 Vol 1. Systems Security Engineering: Considerations for a Multidisciplinary Approach in the Engineering of Trustworthy Secure Systems, November 2016

Souppaya, M., Scarfone, K., Dodson, D. (2022) NIST SP800-218 Secure Software Development Framework (SSDF) Version 1.1 February 2022

UNECE WP29 (2022) World Forum for Harmonization of Vehicle Regulations https://unece.org/wp29-introduction Accessed on 25 November 2022

VDA QMC Working Group 13 / Automotive SIG (2015) Automotive SPICE Version 3.0 July 2015

Wynn, J. (2014) Threat Assessment and Remediation Analysis (TARA). MITRE Whitepaper.

Dragons, Enigmas, Treasure Islands and Zombies: An Iconography of Risks Related to Dead Data

Mike Parsons

AAIP, University of York and SCSC DSIWG Chair

Paul Hampton

CGI UK

Abstract *This talk examines the problems of unused, hidden or legacy data in software systems. For software that has evolved over many years, the purpose and intent behind the data items (e.g. constants in source files, initialisation data, and system configuration values) may be lost in the mists of time, with little or no supporting documentation. Sometimes this data is used within the system, sometimes not, and sometimes it should be used when it is not, and not used when it should be. This talk categorises types of Dead Data, Hidden Data and Change Data and explains how they are identified, named and managed. The naming and metaphorical imagery used is considered relevant, as it helps with identification and awareness in the minds of the developers charged with maintaining complex legacy software systems. There is an analogy with Dead and Deactivated code, a concept used in several standards and guidance including DO-178C.*

© Mike Parsons, Paul Hampton 2023
Published by the Safety-Critical Systems Club. All Rights Reserved

An Overview of IEEE Std 1848-2020: Standard for Techniques and Measurement to Manage Functional Safety and Other Risks with Regards to Electromagnetic Disturbances

Davy Pissoort

KU Leuven

Keith Armstrong

Cherry Clough Consultants

Abstract *This IEEE Standard provides guidance on the assessment and application of techniques and measures that can help reduce the risks associated with the interfering effects of electromagnetic disturbances on digital electronic systems, especially safety- or mission-related systems. When competently selected and applied, a set of such techniques and measures can provide the part of the evidence relevant to EMI required for justifying functional safety decisions and for compliance with functional safety standards (including all applicable parts of IEC 61508 Ed.2:2010 or functional safety standards that are based on IEC 61508). They can also provide part of the evidence relevant to EMI for medical/healthcare systems for which risks are managed in accordance with ISO 14971:2007. This standard supports the adoption of adequate electromagnetic resilience engineering practices throughout the functional safety lifecycle by offering further guidance and practical advice on the application of risk management activities, including the techniques and measures set out in IEC 61000-1-2:2016. While it is primarily intended to be used by those who have responsibilities for functional safety, the methodologies, techniques, and measures it describes can also be used for the reduction of other kinds of risks in any systems that employ electronic technology, such as security risks and non–safety- related risks (for example, risks to the operation of commercial IT systems).*

© Davy Pissoort, Keith Armstrong 2023
Published by the Safety-Critical Systems Club. All Rights Reserved.

A Practical Approach for the automation of product safety case generation in CI Framework

Doria Ramadan

Valeo, CDA, VCORE

Cairo, Egypt

Abstract *The safety element out of context (SEooC) is described by ISO 26262 part 10. It addresses safety-related elements that are not developed in the context of a particular vehicle but rather with assumptions that have to be validated before integration into the final system. It aims at reducing the certification cost through modularization and reuse of element certification evidence. A complete safety case is needed for every release instead of just at the start of production (SOP) within agile product development that uses continuous integration and continuous deployment (CI/CD). So effective approaches to managing the safety cases are needed to fit into the CI framework. In this paper, we provide a practical approach that facilitates implementing the safety-critical applications as fragments of safety elements out of context (SEooC) and automating the merging of modular safety case fragments at the end to build the product line safety case. The approach shows a reduction in development costs. We design it to get fully automated SEooC integration and verification in modern CI/CD frameworks. We get an automated SEooC integration flow starting from integration in the CI process, passing through the verification of assumptions and the configurations, and ending by generating the safety case of SEooC. We build the approach into an embedded testing framework to verify the SEooC integration constraints and ensure the SEooC integrator follows the assumptions mentioned in the safety contract. Finally, the proposed approach leads to having a continuous automated generation of proof of compliance with the safety contract assumptions.*

1 Introduction

We execute the entire 26262 lifecycle in each new project. We deliver it by the end of the project after getting all safety requirements. The SEooC concept allows any supplier to perform life cycle activities out of order, compliant with ISO 26262, and fully covering all needs for the customer to build a safety case before

© Doria Ramadan 2023.
Published by the Safety-Critical Systems Club. All Rights Reserved.

getting all safety requirements from the customers. SEooC development promises cost reduction. Integrating SEooC in a context, requires assumptions of use analysis. Safety contract includes the assumptions of use which have to be followed by SEooC integrator. Integrators may consider reflecting the assumptions of use in the late phase of the project which implies late software changes. We provide SEooC with the left leg of the V cycle and an embedded automated testing framework, which verifies the right leg. Also, it verifies assumptions of use in a way fitting with modern agile product development with continuous integration deployment (CI/CD). By offering a way to verify the SEooC integration and validate the following assumptions of use that are mentioned in the safety contract in an automated way, it leads to having an automated generation of proof of compliance with safety contract assumptions. It ends up constructing a product safety case from safety case fragments. In this paper, the use case depends on microcontroller unit (MCU) self-checks which are part of any safety-related application. They are needed to test the integrity of hardware vital parts and to monitor hardware random faults. So there is a potential to standardize it on the software level to achieve gain from reusability. That's why we developed the SafeTSuite software package as a SEooC. the assumptions on the environment of the SEooC are made without knowing the actual context (item) which the SEooC will be integrated into. We propose a work process combining the use of component-based design, contracts, modular assurance cases, and continuous assessment to enable continuous deployment in the context of product lines.

2 Case study: SafeTSuite Software as a SEooC (Hardware Random Failures Handling)

SafeTSuite is developed as SEooC according to ISO- 26262.10 to provide safety mechanisms implementation of Aurix 2G microcontroller based on Infineon safety manual requirements. The key element of a SEooC is the set of assumptions that are applied during its development such as the ASIL which is B in our study case. It supports running in a multi-core environment whether in master or slave mode. It is designed to support all variants of TC3xx. SafeTSuite includes an embedded automated testing framework to verify the right leg of the V cycle. It is compiled in a CI server such as Jenkins where the ECU (Electronic Control Unit) is connected to the nightly test execution environment. The server checks testing results, generates reports, and notifies the users. The flow is suitable for agile development. Fig.1 shows the SafeTSuite within AUTOSAR considering that it is developed to be run in non-AUTOSAR environments also. SafeSuite was designed following appropriate software standard aspects to fulfill the proposed approach needs.

2.1 SafeTSuite Software Design overview

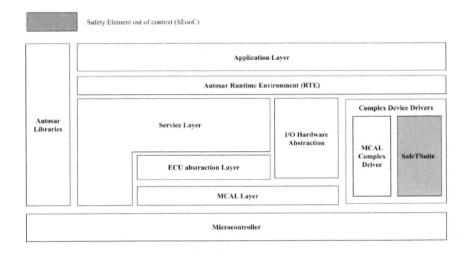

a) Safety library (SafeLib)

It is the software package which has a collection of tests that are required for initiating the integrity of the microcontroller. The software integrator activates and deactivates the tests in the configuration files based on the system requirements once safeTsuite is integrated in a context. SafLib provides two types of tests, tests that run during the startup phase and others to run during the runtime phase. It is also responsible for evaluating test results and reaching the safe state within a fault-tolerant time interval (FTTI).

b) VrfyAou (verify assumptions of use)

This is a software component, which is responsible for assuring the implementation of assumptions of use in real-time. It reports an integration error to the logging feature. So the SEooC integrator will not verify the assumptions in the traditional way which consumes time and sometimes software implementation gaps found in late project milestones. To make the functionality of VrfyAou clearer, we can take an example such as if the AoU mandates to run a memory test for all used memories, it reports an integration error if the cache memory is activated in the software while the SEooC integrator did not configure the test to run on this specific memory. So it ensures requirements are bug free as it acts as automated software design reviewer.

2.2 Design SafeTSuite for Scalability

We designed SafeTSuite to make full use of the scalability of Infineon Tricore devices. The scalability in terms of performance, memory, and packages within the Infineon AURIX family allows for a common safety case across the different devices. Also, it allows single applications to be hosted on smaller devices and on larger devices without the need to modify software architecture or safety strategies.

2.3 Design SafeTSuite for continuous verification

2.3.1 Configurations Verification

There are two lifecycle options described by ISO 26262 to ensure the testing coverage of the configurations. The first option requires verifying the configured software by the customer or by the supplier if it provides a post-release software verification service. The second one requires verifying the SEooC software with a range of admissible configuration data in the SEooC development phase. By applying this option no further verification is needed as this verification activity is already done by the supplier. For SafeTSuite, we combined the advantages of each option to achieve fast, efficient, and low-cost reusability. Sample configurations are verified during the development phase and we provide an automated test framework, which is embedded in the software package so customers can use it to verify any configurations set for any variant belonging in the same microcontroller family such as infineon TC3xx family.

2.3.2 Automated testing framework

The testing framework is embedded in SafeTSuite. It can be compiled by triggering the test command. The output binary is executed in real hardware ECU.

There are many challenges to building the automated testing framework for SafeTSuite:

- SafeTsuite has to be executed in the startup phase and before AUTOSAR communication stack initialization so we cannot use any commutation bus for testing purposes
- Using debuggers is not a good idea as testing has to be run without the intrusiveness of the debugger that may change the SW behaviour in the real scenario. We already experienced issues that did not occur if the debugger was connected.
- The debugger is not available in all environments of system tests.

A Practical Approach for the automation of product safety case generation... 359

Using the universal asynchronous receiver-transmitter (UART) communication protocol helped us to overcome those challenges. UART is initialised in the booting phase. We built the testing framework to log test results on the bus and then automate the UART terminal parsing to generate test reports.

2.3.3 Assumptions Verification

Assumptions are verified by the *VrfyAoU* component. We designed the component to log the test results on the UART bus. The same approach is used for testing configurations, the UART terminal is parsed using scripts to generate the report.

2.3.4 SafeTSuite safety case in CI/CD framework

Recently, the automotive industry is going in a direction of continuous integration and continuous deployment (CI/CD). In CI/CD, the idea is to continuously evolve the product in frequent increments so that all these can be deployed to the end customer. A complete safety case will be needed every release instead of providing it by the start of production (SOP). In our use case, SEooC safety case artifacts including auto generated test reports are consolidated in the CI server. The flow starts when the integrator pushes the SEooC software in the master, the server triggers verification. The embedded testing framework prints the test results in the UART terminal which is parsed using scripts. Scripts generates testing reports and locate it in the safety case folder. SafeTSuite is not allowed to be integrated to the main branch unless the safety contract in this integration fulfils the completeness criteria.

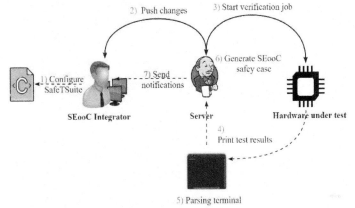

Fig. 1. SEooC Integration flow in CI framework

The proposed approach relies on two major software entities. The first one is 'VrfyAoU' which brings evidence from the right leg of V cycle activities. The second one is an embedded testing framework which is built in an automated environment to ensure full verification without manual efforts.

3 Conclusion and future direction

In this paper, we have proposed a practical approach that shows effectiveness and adaptability to agile development. The approach shown allows a cost reduction of safety critical applications. Software quality is guaranteed as safety cases are built continuously along with the development cycle. We intended to develop our software units as SEooC in an automated CI framework which combines all modular safety case fragments in the product safety case.

References

Rolf Johansson, Håkan Sivencrona. Developing SEooC -Original Concepts and Implications when extending to ADS. CARS 2021 6th International Workshop on Critical Automotive Applications:Robustness & Safety, Sep 2021, Münich, Germany.

'ISO 26262:2018 - Road vehicles -- Functional safety'

J. Birch et al., 'A Layered Model for Structuring Automotive Safety Arguments', in Proceedings of the Tenth European Dependable Computing Conference (EDCC), 2014.

I. Sljivo et al., 'A Method to Generate Reusable Safety Case Fragments from Compositional Safety Analysis', in Software Reuse for Dynamic Systems in the Cloud and Beyond, vol. 8919, I. Schaefer and I. Stamelos,Eds. Cham: Springer International Publishing, 2014, pp. 253–268.

F. Warg et al., 'A Continuous Deployment for Dependable Systems with Continuous Assurance Cases', in Proceedings of the 2019 IEEE International Symposium on Software Reliability Engineering Workshops (ISSREW).

Resilience in Safety Critical Systems – Offshore FRAM

David Slater

Cardiff University

Abstract *When we design systems, it is usual to scrutinise them for any safety issues of concern and ensure the critical components are sufficiently reliable, or to add other systems designed to protect against failures. Such safety-critical systems are in effect adding to the defences against system vulnerability to known scenarios, or "design safety cases". But these additions inevitably make the systems more complex and their control more challenging. Understanding how they behave requires a system-wide model. This model must allow the observation of the possible non-linear, non-predetermined interactions and interdependencies, between subsystems, especially safety-critical ones, which can give rise to unforeseen, emergent, or resonant behaviours. These are often the cause of unexpected and unplanned disturbances in normal operations, which in turn, are normally worked around, but which can occasionally get out of hand and result in significant incidents. Currently the only complex system modelling approach which allows this systematic identification of such resonances, is Hollnagel's Functional Resonance Analysis Method (FRAM). This allows us to pick up safety issues, but also to design-in and evaluate functions to learn from normal operations and to continuously improve the operability of the systems. But the real bonus is, that through this learning, we can utilise the memory, or database, to discern trends in patterns of behaviours, which could enable the anticipation of emerging problems and modify the responses proactively. So, this allows the incorporation of an extra dimension, of not just passive, reactive (imagined?) safety, but proactive operational resilience, for actual complex sociotechnical systems, in the real world. This paper sets out to illustrate this approach, by looking at the safety-critical systems in the Macondo Well Blowout accident.*

© David Slater 2023.
Published by the Safety-Critical Systems Club. All Rights Reserved.

AUTHOR INDEX

Paul Albertella	19
Tom Anderson	21
Keith Armstrong	353
Stephen Bull	57
Simon Burton	297
Laure Buysse	83
Lucia Capogna	57
Carmen Cârlan	105
James Catmur	107
Aditya Deshpande	339
Rob Davies	57
Dai Davis	141
Alastair Faulkner	143, 161
Jane Fenn	163
Reza Ghabcheloo	281
Stephen Gill	57
Paul Hampton	189, 351
Tom Hughes	213
James Inge	215, 233
Craig Innes	247
Andrew Ireland	247
Gill Kernick	265
Kevin King	107
Aimée M.R. de Koning	281
Yuhui Lin	247
John A McDermid	297, 317
Jamie McFadden	337
Catherine Menon	335
Mark Nicholson	143, 161, 163
Robert Oates	339
Ganesh Pai	163
Mike Parsons	21, 107, 317, 351
Davy Pissoort	83, 353
Zoe Porter	297
Austen Rainer	335
Doria Ramadan	355
Subramanian Ramamoorthy	247
Roger Rivett	21
David Slater	361

Fathi Tarada...107
Jens Vankeirsbilck..83
Dries Vanoost..83
Simon Whiteley...83
Michael Wilkinson...163
Phil Williams...233

Printed in Great Britain
by Amazon